THE ECOMMERCE GROWTH PLAYBOOK

CUT FRICTION, KILL TECH DEBT, AND SCALE YOUR MID-MARKET BRAND—FAST

JOSHUA S. WARREN

CREATUITY PRESS

First edition

ISBN 979-8-9995287-0-4 (paperback)

ISBN 979-8-9995287-1-1 (hardback)

ISBN 979-8-9995287-2-8 (e-book)

Published by **Creatuity Press**

1255 W 15th St, Suite 470, Plano, TX 75075 | www.creatuity.com

Printed in the United States of America

❀ Formatted with Vellum

To Jenna, Anna, and Luke, who make every milestone matter.

CONTENTS

WELCOME & HOW TO USE THIS BOOK

As a voracious reader who usually feels like he must read a book cover-to-cover, one chapter at a time, I want to tell you now - you don't have to do that here!

If you're a cover-to-cover person, awesome - you'll get a lot out of this book. But if you're like most ecommerce managers and executives I coach, you're too busy to read all 100,000 words in this book. I get it, and I'm not even offended.

So, if that's you - feel free to skip around. Let me walk you through how the book is laid out so you can get the most out of it.

Oh, and grab a pen (or make lots of Kindle highlights!) Every dog-eared page, scribbled margin note, and highlight you add here turns this book from 100,000 words of theory into a custom growth plan for your brand.

THREE BOOKS IN ONE

I numbered the chapters intentionally, chapters 2.1 thru 2.4 make up the first arc on customer experience, then 3.1 through 3.4 are the second arc, on technology. Finally, chapters 4.1 through 4.4 cover leadership and ecommerce team structure.

Here's what you'll find in each arc:

Customer Experience – speed, search, checkout, omnichannel.

Ecommerce Technology – tech debt, data plumbing, composable stacks, site speed.

Leadership & Org – scorecards, culture, vendor control, change when everyone's fried.

Any one of those could stand alone, but mid-market operators don't have the luxury of "read UX now, tech next quarter." Problems land all at once. So I bundled everything you'll need for the next 12–18 months into one field guide.

Skip Around—Guilt-Free

If page speed is on fire, jump straight to **Chapter 3.4** and run the 7-day leak test tonight. Wrestling with agencies? Flip to **Chapter 4.4** and copy the red-flag checklist into your next weekly call. Each chapter is self-contained:

• A blunt reality check (why this topic pays the bills).

• Benchmarks and hero metrics (know if you're winning).

• Case studies you can steal.

• A 7-day or 90-day action plan depending on the complexity of the topic.

Read, deploy, measure, move on. Come back later when the next fire flares.

How the Pieces Fit

Still craving a linear path? Here's the high-altitude map:

Chapter 1.1 – See the silent revenue tax you're already paying.

Chapter 2.1 through 2.4 – Scrub friction and engineer experiences that convert.

Chapter 3.1 through 3.4 – Harden the engine under the hood so growth doesn't snap.

Chapter 4.1 through 4.4 – Give the right people the right score-boards and get out of their way.

Chapter 5.1 - 90-Day Action Playbook – Tie it all together, quarter by quarter.

Follow that arc once if you're new to the chaos. Afterward, treat the table of contents like a toolbox.

METRICS FIRST, EGO LAST

Every chapter hands you KPIs and a target. Record your baseline in the margin before you try a single tactic. Real data beats opinions—including mine.

PLAYBOOK CADENCE

The playbooks stack like this:
- **Weekly wins** – tiny, shippable tasks.
- **30-, 60-, 90-day milestones** – traction you can brag about in the next board meeting.
- **Quarterly review ritual** – lock gains, pick the next metric, repeat.

If a task looks too big, slice it in half until it fits inside a five-day sprint. Momentum matters more than perfect sequencing.

WHAT YOU'LL NEED

- One empowered Product Owner (yes, someone who can say "ship it").
- A living dashboard—Google Sheet, Looker Studio, cave wall, doesn't matter. And yes, I worked with an ecommerce team once who were relocated into what might as well have been a cave - it was a windowless room with plywood walls.
- The courage to delete more than you add.

HOW LONG WILL THIS TAKE?

Brands that attack one issue at a time see wins inside two weeks and step-change growth inside six months. Brands that "plan to plan" are still rewriting strategy decks next year. Your call.

MY ASK

If a chapter moves a metric, email me the before-and-after screenshot. I'll send back a video clip with your next improvement idea. Deal? My contact info is at the end of the book.

LET'S GET MOVING

Flip to the chapter hurting the most, circle one task, and start a timer for ten minutes. When it dings, you should be deeper than "thinking about it." Shipping beats studying. Every. Single. Time.

See you on the other side.

1.1: WHY GOOD ENOUGH IS A SILENT REVENUE TAX

I was on a discovery call with a new client when they said something that made me wince. "Our site is fine; it's good enough for now," the ecommerce director told me. In that moment, I knew "good enough" was costing them more than they realized. As we dug deeper, we saw that on the surface their site was indeed okay. They had a decent conversion rate and no major outages. But the cracks were there. Pages were loading a second or two slower than at their competitors. The mobile checkout was clunky. And their search function was outdated and frustrating shoppers. These weren't setting off alarm bells for the team because nothing was on fire. Yet every one of those "minor" issues was quietly chipping away at their revenue. Shoppers who encountered a slow page or a mediocre experience didn't complain—they just left. The hidden cost of settling for "good enough" was an unseen tax on their growth.

STUCK IN A GROWTH PLATEAU DESPITE SOARING AD SPEND

• • •

That discovery call reminded me of another retailer I worked with a while back. Let's call them TrendHouse. TrendHouse had hit a growth plateau at around $20 million in annual online sales. Year after year, they pumped more money into Facebook and Google ads. They increased spend by 10% one quarter, then another 15% the next. Their customer acquisition cost kept rising, but revenue stayed flat. They were basically paying more just to *tread water*. Why? Because while their marketing spend ballooned, their site experience stayed the same. Their homepage, product pages, checkout flow—everything was adequate but uninspired. They figured the funnel was "good enough" and that more ads would bring growth. Instead, all that ad spend was pouring water into a leaky bucket.

When we finally overhauled their approach—optimizing page load times, simplifying navigation, and adding personalized recommendations—their conversion rate jumped and the plateau broke. The lesson from TrendHouse was clear: you can't coast on a mediocre site and expect paid ads to save you. Good enough isn't good enough, not if you want to grow.

MID-MARKET ECOMMERCE: CAUGHT BETWEEN AMAZON AND HARD PLACES

Most of the clients I coach are what I call mid-market ecommerce companies—typically $10 million to $750 million in annual online revenue. If that sounds like you, you know the unique squeeze you're in. You're not a scrappy startup, but you're also not Amazon or Walmart. Customer expectations, however, don't care about your size. Shoppers have been trained by the Amazon experience to expect blazing-fast load times, 24/7 convenience, and seamless everything. If your site lags, even by a bit, people notice. In fact, about 78% of shoppers will abandon a site that loads too slowly. Nearly as many won't return after a bad user experience.

Mid-market teams also face resource constraints. You likely don't have a 50-person IT department or a squadron of CX analysts on staff.

Hiring is tight. You need folks who can wear multiple hats because you can't afford a specialist for every niche. Meanwhile, you're probably juggling a tech stack of a dozen or more SaaS tools. There's one for email marketing, one for reviews, another for inventory, plus your ecommerce platform, analytics, personalization, search... the list goes on. The average mid-market brand uses 14+ different software services to run their store. It's a lot to manage. If those tools don't play nicely together, your team ends up playing tech support instead of focusing on growth.

On top of that, you're under pressure to deliver big results quickly. Investors or executives set aggressive targets. But unlike the enterprise giants, you can't simply throw money at problems. You have to be smarter and scrappier. It's like being asked to win a race while driving a sensible sedan against someone else's racecar. The good news is, a well-tuned sedan can win. You just have to optimize the engine and drive efficiently.

A Tale of Two Approaches: Slow and Steady vs. Ship and Learn

Let me contrast two companies I've seen, to highlight a key mindset difference. Company A (we'll keep them anonymous) took the "if-it-ain't-broke" approach. They would do a major site overhaul once a year at most. The rest of the time, they avoided changes to keep things stable. Deployments were infrequent and stressful — big batches of updates with weeks of QA. Sure, they had fewer releases, but each one was a giant leap with higher risk. And in between those big launches, their site stayed stagnant while the market moved on.

Company B, on the other hand, believed in continuous improvement. For example, Worthington Direct—a B2B furniture retailer I've worked with—took this iterative approach. They embraced a deployment cadence of smaller updates every few weeks. They treated their ecommerce site not as a one-and-done project, but as an ongoing product that gets a little better every month. Did every tweak deliver a huge win? No, of course not. But by shipping regularly, they could test ideas and fix small issues before they became big ones. They could also respond to customer feedback in near real-time. Over the course of a year, those incremental gains compounded. Meanwhile, Company A (with their once-a-year big redesign) found themselves losing customers to more agile competitors.

The moral: deployment frequency matters. If you're iterating continuously, you're learning continuously. If you're only updating your site a couple of times a year, you're stuck in a time capsule for the other 10 months. In the fast-paced world of ecommerce, standing still means falling behind.

THE THREE LEVERS OF BREAKOUT GROWTH

So how do you break out of "good enough" and plateau problems? This book is built around three core levers that any mid-market ecommerce leader can pull to accelerate growth. Think of these as the pillars of your Growth Playbook:

• **Customer Obsession:** Relentlessly focus on your customer's experience and needs. This isn't just lip service about being "customer-centric"—it means constantly asking, *"What does the customer really want and where are we letting them down?"* It could be site speed, product selection, post-purchase follow-up, or all of the above. For example, **Azio Beauty** (a skincare brand) realized their customers often ran out of their products and forgot to reorder. They set up a predictive replenishment program to remind customers when it was time to re-up. As a result, Azio saw an **18% jump in repeat orders**. That's customer obsession in action: deeply understanding the customer's life cycle and adding value at just the right moment.

• **Scalable Engine:** This is about the technology and processes that power your business. Your site, your backend systems, your automation workflows—all need to scale smoothly as you grow. A scalable engine is efficient and built for speed and reliability. It's also about operational excellence. Do you have a tech stack that can handle surges in traffic? Are your pages optimized to load in a blink? When you send out a promotion, can your inventory and fulfillment keep up? Back-end improvements aren't always sexy, but they directly impact revenue. One mid-market brand I worked with shaved their average page load time from 4 seconds down to 2 seconds. Their conversion rate went up noticeably because fewer customers bounced. A scalable engine lays the groundwork so that when you pour on the gas (marketing), the car can actually accelerate without breaking down.

• **Leadership & Organizational Design:** Even with the best tools and customer focus, you need the right people and structure to execute. This lever is about how you build and empower your team. Do you have clear ownership of your ecommerce KPIs? Is someone waking up every day feeling responsible for conversion rate or repeat purchase rate? Companies that win have leaders who champion a culture of experimentation and accountability. A great example is a B2B seller of chemicals I've worked with. They appointed a dedicated ecommerce product owner and gave her the authority to make quick improvements without a bureaucratic gauntlet. That empowerment led to a surge in new features and optimizations on their site within months. These were things that previously would have been stuck in committee. When your team has the right structure and the freedom to act, they will drive growth in ways top-down management never could.

Throughout this book, we'll dive deep into each of these levers in dedicated chapters. But here's the key: these levers work together. Customer Obsession tells you *what* needs to improve from the shopper's perspective. Scalable Engine gives you the *how* to actually deliver it reliably and fast. Leadership & Org Design ensures you have the who and culture to keep the whole machine running and improving.

• • •

WHICH TYPE OF BUSINESS ARE YOU? (FIND YOUR ARCHETYPE)

Advice isn't one-size-fits-all. A tactic that works for a fashion D2C brand might not make sense for a B2B industrial supplier. Over the years, I've noticed mid-market ecommerce companies generally fall into one of three archetypes. Knowing which bucket you're in can help you focus on the most relevant growth levers:

- **Side-Car B2B:** You might be a manufacturer or wholesaler that traditionally sold through sales reps, catalogs, or distributors, and now you're adding ecommerce as a *side-car* to your core business. In this archetype, online sales might still be a smaller slice of revenue, but it's growing. Your challenge is often integrating online channels with your traditional sales model. You also need to get buy-in for digital from a historically offline organization. (Think of a B2B parts supplier launching an online catalog to supplement their phone orders.)

- **Omnichannel Retail:** You operate both online and brick-and-mortar stores (and maybe other channels like catalogs or marketplaces). For you, ecommerce is one piece of a larger puzzle. Your customers expect a seamless experience whether they're in-store or on your site. You have to worry about things like in-store pickup, consistent inventory across channels, and unified marketing. A common challenge here is breaking down silos—making sure your online team and your store operations are in sync and not competing for resources or credit.

- **Pure-Play Digital:** You're an online-only business. Maybe you're a direct-to-consumer brand born on Shopify, or an online B2B vendor with no physical storefront. All your revenue flows through the digital channel. The upside is you can be laser-focused on ecommerce. The challenge is that everything about growth falls on your digital execution. If your website falters, your sales stop—period.

Pure-plays often live and die by their ability to acquire customers online efficiently and keep them coming back. There's no offline safety net at all.

You might see a bit of yourself in more than one archetype, and that's okay. The point is to orient yourself: which description sounds most like your situation? Throughout the book, I'll call out tips or watch-outs specific to each archetype when relevant, so you can adapt the playbook to your context.

How to Use This Playbook

I want this book to be as practical as possible. Think of it like a personal coaching session with me, but in book form. Each chapter after this will focus on a specific area (often tied to one of the three levers) and will usually include a checklist or a set of action items. Here's how I suggest you use it:

1 Pick a Focus: Start with the chapter that addresses your biggest pain point or opportunity. Maybe your site is slow and you know it (the Scalable Engine chapter will be your friend), or maybe you have lots of traffic but low repeat business (jump to Customer Obsession and retention strategies).

2 Run One Checklist Item: At the end of each chapter, you'll find a checklist or a set of quick wins. Pick one item that you can implement *this week*. Not all ten, not "boil the ocean"—just one.

3 Ship One Change: Get that change live on your site or in your process as fast as you can. Speed matters. If it's a bigger project, break it down. The goal is to **build momentum** through action, not endless planning.

4 Measure the Impact: Once the change is live, watch your metrics. Did the needle move? Sometimes it will, sometimes it won't, but you'll always learn something. And if it *did* work, you just earned a win you can point to.

5 Rinse and Repeat: Then move on to the next item, or the next chapter, and do it again. Growth is the sum of dozens of small improvements, plus the occasional big win. The more shots on goal you take, the more chances you have to score.

By approaching the book this way, you'll turn ideas into results. It's not about reading theory—it's about **putting the playbook into play**.

BENCHMARK YOUR STARTING POINT

Before you dive into making changes, I have a challenge for you: know your baseline. Take a moment to write down a few key metrics as they stand today. This will be crucial for measuring your improvement as you apply tactics from this playbook:

- **Conversion Rate (CR):** Your overall website conversion rate— what percentage of visitors buy something. If you have separate numbers for desktop vs. mobile, note those too.
- **Average Order Value (AOV):** How much does the average customer spend per order? This might be one of the easiest levers to pull once you start optimizing, so log the current number.
- **Repeat Purchase Rate:** What percentage of your customers have purchased more than once? In other words, are you getting second and third orders, or is it mostly one-and-done? This is a good indicator of loyalty and how well you're turning first-time buyers into loyal customers.
- **Deployment Frequency:** How often are you updating your website or app? Be honest. Is it daily, weekly, monthly, quarterly? You don't need an exact count, just a general sense of your pace. This number is a proxy for how agile your team is. If it's "a few big updates a year," you've got a huge opportunity to improve.

Write these down somewhere visible. These are your starting stats —your baseline. As you iterate through changes, refer back to this list. You'll start seeing these numbers move in the right direction, and that's incredibly motivating. If something isn't moving that you expected to, that's a signal too. Maybe you need to dig deeper or try a different tactic.

· · ·

No Coasting Allowed

The last thought I want to leave you with in this introduction is a mindset shift. Coasting is a tax, not a strategy. It's easy to think that if you just maintain the status quo—keep the site running, keep the lights on—you'll be fine. But in ecommerce, standing still is really just slowly moving backwards. Every month your site is "okay" but not improving, you're paying a tax. That tax comes in the form of missed revenue, lost customers, and widening competitive gaps.

The companies that win aren't necessarily those with the deepest pockets. They're the ones with the mindset that growth is a constant pursuit. They know that every day is an opportunity to get better, even if just by 1%. And they know that not improving means letting things deteriorate, because the world around you isn't waiting.

So as we kick off this playbook, I challenge you to adopt that mindset. Don't accept "good enough" from your team, your site, or yourself. There's too much opportunity out there to waste by coasting.

Alright, that's enough pep talk. Let's dive in and start turning "good enough" into great.

2.1: THE COST OF FRICTION

t's 8:59 PM on Cyber Monday. I'm on a video call with a frantic e-commerce CEO watching live analytics. Her team spent millions driving traffic to a holiday sale, but conversions are flat. Users are hitting the site, adding items to cart... then vanishing. The culprit is obvious: the mobile page takes five full seconds to load. By the time it appears, the customer is gone. In those few seconds of delay, we literally watch thousands of dollars slip away in real time. This is the brutal cost of friction – the invisible killer of e-commerce success. I've seen it too many times: slow pages, clunky checkouts, confusing returns, or a disconnected store experience. Each little hassle acts like sandpaper on your customer's goodwill, quietly grinding your revenue down. Today, I'm not here to sympathize; I'm here to challenge you to hunt down and obliterate these friction points. Your business depends on it.

What Do I Mean by "Friction"?

In e-commerce, **friction** is anything that makes a purchase harder than it needs to be. It's the drag in your customer's journey – the extra clicks, the slow loads, the doubts about trust, the hoops to jump through at checkout. Friction is the silent revenue killer you probably aren't measuring on your balance sheet. But I guarantee it's eroding your sales from the shadows. Think of friction as the **"experience tax"** you unintentionally charge your customers. Every inconvenience – no matter how small – is a tax that some customers refuse to pay. They leave instead. And when they leave, they take their money with them.

Let's define it clearly: friction is anything that causes hesitation or dropout on the way to conversion. That could be a page that lags, a form that's too long, a return policy that's a hassle, or even a skeptical feeling that your payment process isn't secure. In over two decades of e-commerce work, I've learned to spot these friction hotspots quickly. And once you spot them, you can eliminate them – if you're willing to confront some hard truths about your store's experience.

WHERE FRICTION HIDES: THE SIX HOTSPOTS

Friction can lurk anywhere in your commerce ecosystem, but six common hotspots account for most of the damage. Let's expose each one:

• **Mobile Speed.** This is the era of mobile-first customers, yet too many sites still load like it's 2005. If your mobile site is slow, you are literally driving away money. A slow storefront hurts conversions and revenue. Customers won't wait; a few seconds of delay and they're gone to a competitor. One of our Commerce Today episodes bluntly noted that ignoring mobile optimization is an e-commerce sin. Speed isn't a luxury; it's a basic requirement. Check your load times on a 4G connection – if it's not near-instant, it's friction.

• **Checkout Process.** Look at your checkout through a customer's eyes. Is it quick, or is it an interrogation? Every extra field, every forced account signup, every surprise fee is friction that *kills* conversions. The industry knows this: nearly three-quarters of shopping carts are abandoned on average, largely due to checkout friction. Think about that – you've done the hard work to get a customer all the way to checkout, and 70% of those orders evaporate due to some annoyance or obstacle. That is a colossal leak in your revenue bucket. Simplify checkout to the absolute essentials. I often coach clients to remove half the fields (yes, half!) and watch conversion rates climb.

• **Returns Experience.** Many merchants treat returns as an afterthought or a necessary evil. Big mistake. A clunky returns process is major friction *and* a trust killer. Customers expect returns to be smooth and painless – a seamless extension of the purchase. And if you think strict policies will reduce returns, think again. A Retail Dive report found 91% of retailers saw returns growing faster than revenue, with roughly 30% of online orders coming back. That's the reality. If you make returns hard, those customers won't come back. On the flip side, a frictionless returns process (print the label, easy drop-off, speedy refund) builds confidence that encourages shoppers to buy in the first place. Don't fight the returns tidal wave – ride it by making it effortless.

• **Channel Flow (Omnichannel Integration).** Here's a test: order a product on your site for in-store pickup. Is the item actually ready when you show up? Or are you met with blank stares because the online and store systems don't talk? Too often, online and offline channels behave like isolated "ecommerce islands" rather than a unified experience. I call this *channel friction*. Customers see your brand as one whole; any internal silo is not their problem – it's yours. We tackled this in a Commerce Today deep dive on omnichannel: bridging online and physical stores requires nailing operational and technical integration. If you don't, you get scenarios like a customer buying online and the store never getting the memo – a surefire way to lose a sale and goodwill. Every handoff between channels (store to web, web to mobile, social to cart) needs to be seamless. If it's not, that's friction you must fix.

• **Payment & Trust.** The moment of payment is where a customer's trust in you either solidifies or shatters. Any hint of insecurity or inconvenience here is deadly. Does your customer hesitate, wondering if your site is legit or their card info is safe? That's friction. Different markets have different trust factors – in some places, customers still prefer physical cash because they don't trust online payments. In one podcast episode we explored strategies to cultivate unwavering consumer trust in e-commerce. Clear information, security badges, trusted payment options (Apple Pay, PayPal, etc.), and a smooth, reassuring payment page all matter. Also, don't force a particular payment method. If a shopper wants to use PayPal or a "buy now, pay later" service and you don't offer it, that's unnecessary friction. Give people options they trust. Trust isn't a soft metric – it directly impacts whether the transaction happens at all.

• **In-Store Digital Friction.** E-commerce doesn't get all the blame. Physical retail has plenty of friction too, especially as stores become more digital. A great example is Starbucks' digital ordering fiasco a few years back. Starbucks rolled out an ultra-efficient mobile ordering system that was too successful – customers flooded in, mobile orders piled up, and the in-store experience went out the window. The company became so fixated on digital efficiency and sales volume that they neglected the community-driven coffee shop vibe that made their brand special. The result? Frustrated baristas, crowding, confusion – in short, friction. Under CEO Brian Niccol, Starbucks recognized the mistake and rebalanced tech with experience. The lesson: whether it's self-checkout kiosks, mobile order pickup, or any in-store tech, it must enhance the experience, not just efficiency. And if you're an online-only brand doing pop-up shops or events, remember that people expect the same seamless ease in person as online. Long lines, poor in-store tech, or staff who can't find online order info – that's friction that can taint your brand across all channels.

Take a hard look at these six hotspots in your own business. I guarantee at least one (likely several) are costing you more than you think. And that brings us to the dollars-and-cents part: just how much is friction draining from your bottom line?

The Bottom Line: Quantifying the Cost

Let's put this in perspective. Friction is expensive. In fact, it's outrageously expensive because it scales with your growth. The more traffic you drive, the more potential customers friction can repel. The tricky part is that many businesses don't track it directly. You won't see a line item on your P&L for "lost sales due to slow mobile site" – but it's very real.

Need proof? Amazon (the king of reducing friction) once famously found that every 100 milliseconds of page load delay cost them 1% of sales. Read that again: a tenth of a second = 1% of revenue, which was estimated around $107 million back in 2006. At Amazon's scale today, that would be nearly $4 billion in potential sales lost for just a blink of an eye in speed. If that stat doesn't make you slam the accelerator on site performance, nothing will. And you're not Amazon – your customers are even less likely to wait around.

It's not just Amazon. Industry-wide, the aggregate cost of friction shows up in metrics like cart abandonment (billions in lost sales from those 70% of carts that never convert), return rates (as we saw, 30% of orders boomeranging back – a huge cost center), and customer lifetime value erosion (each bad experience means that customer probably won't return). When 91% of retailers report returns growing faster than revenue, it signals a profitability crisis: friction in the post-purchase experience is eating margins alive.

Friction also has indirect costs that are just as damaging. It kills your marketing ROI – you might be paying handsomely to bring in visitors via Google or Facebook ads, but every point of friction means a chunk of that spend is wasted on people who click and bail. It damages your brand. Customers might not tell you when they experience friction; they just quietly choose a competitor next time. You'll never know why – only that your customer acquisition costs keep rising while repeat purchase rates stay flat.

In short, friction is a revenue leak *and* a profit killer. When you remove friction, everything moves up and to the right: conversion rates improve, average order values often increase (because happy, confident customers buy more), return rates can even drop (for example, better product info and sizing tools mean fewer unhappy returns). The cost of friction is the loss of potential – all the sales you're not making and the customers you're not keeping. So it's critical to understand where *your* biggest leaks are. Which brings us to an honest look in the mirror: based on what kind of business you are, where are your friction risks?

DIFFERENT BUSINESSES, DIFFERENT FRICTION RISKS

Not all e-commerce operations are the same, and neither are their friction hot zones. I've worked with scrappy one-person startups and multi-billion-dollar retail giants; each has its own "friction profile." Let's talk about a few archetypes. See which one sounds like you – and take note of the friction risks that could be lurking:

• **The Lean Start-Up:** You're small, fast, and likely focused on getting the product and marketing right. The danger here is *underesti-*

mating UX details. Maybe you're on a basic Shopify template that hasn't been optimized. Mobile speed might be poor, or checkout is the generic default (not great). Early adopters might forgive some hiccups, but as you try to scale, those little friction points will be brutal. *Risk:* assuming that a decent out-of-the-box experience is good enough. It's not – polish your site now, before growth amplifies the friction.

• **The Scaling Mid-Market Merchant:** You've found product-market fit and things are growing. Perhaps you're doing $5M to $50M in online sales. Commonly, you start encountering process frictions – legacy plug-ins, bolted-on systems, maybe an ERP integration that doesn't sync in real-time. Your team is juggling multiple tools that don't talk to each other. This leads to errors and delays (e.g. overselling stock because inventory sync lagged). Risk: operational friction behind the scenes spilling over to customers (like late shipments or out-of-stock orders). You need to audit both your tech stack and processes and streamline them now. Complexity is the enemy of smooth customer experiences.

• **The Enterprise or Legacy Retailer:** You're big. You might be a traditional retail brand with e-commerce bolted on, or an e-com giant grown through 10 years of iterations. Your biggest friction risk is technical debt and inertia. I guarantee you have some convoluted checkout flow or old database that makes everything slower. Maybe your site search is awful because it's tied to an ancient platform. Or your mobile site is actually a separate m-dot site (hello, 2010 called). *Risk:* assuming that what got you here will get you there. It won't. Giants fall hard when they ignore friction – customers don't care how famous you are if your experience is a pain. You must actively seek out and kill outdated elements in your experience.

• **The Omnichannel Brick-and-Click:** You run physical stores and an online channel. For you, the nightmare is disjointed customer journeys – the right hand (store) not knowing what the left hand (online) is doing. If a loyal in-store customer can't access their purchase history or loyalty points on your website, that's friction. If your site says something is in stock for pickup but the store inventory is wrong, that's friction. *Risk:* customer confusion and frustration with inconsistent information. You need a single view of the customer and inventory.

Invest in the integrations and don't let your channels operate as silos. The best omnichannel retailers treat friction as the enemy of loyalty – because it is.

• **The Marketplace-Dependent Seller:** Maybe you sell primarily through Amazon, Etsy, or another marketplace, with your own site as a secondary channel. Your friction risks are twofold: the friction you control on your site (which often gets neglected since marketplace is priority) and the friction of dependence on another platform's rules. If Amazon changes something (as they often do), your customers might face unexpected friction (like a new return policy) and blame you. Risk: lack of ownership. I'll be blunt – relying solely on marketplaces can hide your friction problems until it's too late. Even if most sales are off-site, optimize your own channels like your business depends on it. It just might if the marketplace rug gets pulled out.

These are broad strokes, but I want you to identify with one (or a mix) of these archetypes and be brutally honest: *where are you vulnerable?* Each archetype has blind spots. Recognizing yours is the first step to eliminating friction at the root. And that leads directly into having a system to find and fix friction systematically. Enter my five-step "friction audit ladder."

THE 5-STEP FRICTION AUDIT LADDER

Eliminating friction isn't a one-and-done task – it's an ongoing discipline. Here's a five-step ladder I use with clients to methodically audit and annihilate friction. Think of this as climbing up out of the quicksand, one solid rung at a time:

Step 1: Walk in the Customer's Shoes – End to End. This sounds obvious, but when's the last time you personally went through your entire shopping journey like a first-time customer? Do it *today*. Browse your site on a phone with fresh eyes (better yet, watch someone else do it while you observe quietly). Add a product, checkout, attempt a return, contact support. Note every annoyance or confusion, no matter how small. At this stage, you are simply gathering raw firsthand experience. I can't tell you how many executives have never actually done a full run-through of their own site's purchase flow. Don't delegate this. Feel what your customer feels.

Step 2: Measure the Misery. Now quantify what you found. For every friction point, attach a metric. Slow mobile load? Pull your Google Analytics – what's the bounce rate on mobile landing pages? How many seconds to interactive? Checkout abandonment? There's data for that – your drop-off rate at each funnel step. Customer support contacts? Measure complaint volume about issues (late delivery, confusing returns, etc.). Put real numbers to the pain. For example, if 60% of visitors bounce after viewing one page on mobile, that's a 60% friction-induced loss at the top of funnel. As the saying goes: if you can measure it, you can manage it. So measure it.

Step 3: Prioritize the Biggest Leaks. Not all friction is equal. Some issues are mild annoyance; others are full stop "I'm outta here" deal-breakers. Focus on the latter first. Using your data, identify the friction points costing the most money right now. Is it the 5-second mobile load time (huge drop-offs)? Is it a particular form field causing 80% of abandonment? Maybe your international shipping is so expensive or slow that foreign customers never convert. Rank these by impact. A simple way: estimate the dollars each issue might be costing per month (even a rough guess based on conversion impacts). Your #1 friction issue is the one with the biggest dollar sign next to it. That's your prime target.

Step 4: Fix, Simplify, or Bust Through. Now roll up your sleeves and solve the problems. This might mean asking your developers to optimize images and scripts to boost mobile speed. It might mean removing fields from checkout, enabling auto-fill, or integrating Apple/Google Pay to simplify payment. Perhaps you need a better returns portal or a clearer policy that's customer-friendly. For channel issues, maybe it's investing in an order management system so online and store inventory stay in sync in real time. Each friction point will have its own solution path: the mantra is simplify, simplify, simplify. Remove steps, automate something, give more transparency – whatever it takes so that the customer glides through. Set aggressive deadlines. I often propose "7-day sprints" per major friction item – most can be fixed or at least noticeably improved in a week if you truly prioritize it.

Step 5: Ladder Up to Prevent Future Friction. With immediate fires quenched, the final rung is building habits and systems to catch friction before it starts. Think in terms of ongoing audits – maybe a quarterly friction review where you or mystery shoppers audit the experience anew. Bake friction metrics into your regular reports: page speeds, abandonment rates, customer satisfaction scores. Train your team to spot friction: frontline employees should feel empowered to report "hey, we keep getting calls about X, maybe we should fix that." The idea is to create a culture that is allergic to friction. When your whole team is constantly smoothing the path for the customer, you've reached the summit of this ladder. Friction will never be zero, but it can be minimized as a constant practice.

Climbing this audit ladder takes humility to admit where you've let customers down and discipline to keep at it. But I promise, every rung is worth it. Each step you take will reclaim revenue that you didn't even realize was slipping away. I've seen businesses transform their growth trajectory just by executing this process ruthlessly for a few months.

Throughout these steps, keep recalling the stories and stats we discussed. Remember the CEO watching those sales vaporize on slow mobile pages, or Amazon's million-dollar milliseconds, or Starbucks learning that efficiency at the expense of experience backfires. These aren't just stories – they're warnings and motivation. Now, let's translate all of this into action for you, right now.

Your 7-Day Quick-Win Plan

I'm a big believer in speed. So I'm challenging you to tackle friction *this week*. Not next quarter, not when "things are less busy." Now. Here's a 7-day quick-win plan to kickstart your friction annihilation. One task per day – do this, and in a week you'll already see results:

Day 1: Speed Audit and Boost. Measure your mobile and desktop load times (use a tool like Google PageSpeed Insights). Identify the biggest speed hogs (images, scripts, server latency). Implement at least one fix *today* – compress images, enable a CDN, whatever it is. Even a small improvement (500ms) is meaningful. Aim to shave off as much load time as you can within 24 hours. Instant gratification: you'll likely see conversion uptick within days of speeding up.

Day 2: Checkout Cleanup. Go through your checkout flow and ruthlessly cut or simplify. Remove any optional fields or steps. Enable guest checkout if it's not on. Test an order on mobile – is it thumb-

friendly? Today, implement at least two changes (for example, eliminate an info field and enable address auto-complete). Your motto:
streamline, then streamline more. Fewer clicks, fewer keystrokes, fewer
surprises.

Day 3: Trust Tune-Up. Review your site through the lens of a skeptical customer. Do you clearly show shipping costs and delivery times
up front? Is your return policy easy to find and written in plain
language? Address one major trust factor today: it could be as simple
as adding customer reviews on product pages or an FAQ answering
"Is my payment secure?". Make your site feel safe and reliable in
every way.

Day 4: Return Revolution. Take a hard look at your returns
process. Today, do one thing to remove friction: maybe generate a
simple return label PDF template, or integrate a returns management
app, or at least rewrite the policy to sound generous rather than
begrudging. If you can afford it, consider offering free returns (at least
as a test) – it can pay for itself in increased sales. The goal for today:
make returning an item almost as easy as buying it.

Day 5: Omnichannel Sync-Up. If you have physical locations or
multiple sales channels, spend today checking their integration. Is your
inventory syncing properly across channels? Test a BOPIS order or a
curbside pickup. Identify any disconnect. Quick win idea: implement
an "order ready" notification if you don't have one, so customers know
exactly when to come and won't wait around. Unify the experience –
customers should feel like your online and offline are one cohesive
system.

Day 6: Team Huddle on Friction. Gather your team (even if it's just
a couple people or different departments) and share the mission.
Explain the key friction points you're attacking and why it matters.
Invite their input – they might know pain points you aren't aware of.
Establish a simple channel for them to report customer pain (a Slack
channel or weekly meeting). By close of Day 6, every team member
should know that eliminating friction is a top priority and feel empowered to suggest improvements continuously.

Day 7: Customer Communication. This one might surprise you:
reach out to a few customers (personally, if possible) who recently

abandoned carts or had returns. Ask them (nicely) what went wrong or what could've been better. This isn't a marketing survey – it's a genuine attempt to learn. The insights you get could reveal a friction point you never considered. As a bonus, those customers will appreciate that you cared enough to ask. By end of day, compile what you learned and add it to your friction hit list.

Follow this seven-day plan and you will have notched tangible wins in a week. More importantly, you'll have momentum. Friction, once revealed and attacked, tends to keep dissolving under that kind of pressure and focus.

Finally, let's zoom out. Eliminating friction isn't just a one-week project – it's a new mindset for your business. You've seen how costly it is, but also how addressable it is with the right approach. No excuses going forward. Make a commitment that friction is the enemy you'll always be hunting. Your customers may never explicitly thank you for a smoother experience, but they'll show it with their wallets and their loyalty.

And guess what? Reducing friction often has a side effect: it forces you to improve other aspects of your business (processes, technology, team coordination) in the process. It's a rising tide that lifts all ships.

You've got seven days of quick wins to conquer and a lifetime of continuous improvement ahead. In the next chapter, we'll build on this momentum and tackle the next big lever of your commerce success. So catch your breath – briefly – and get ready to climb the next rung in our journey. The fight against friction has primed you for what's coming, and I promise you, it's worth it. Let's keep going, onward and upward, because the cost of friction is one expense we refuse to pay anymore.

2.2: EXPERIENCE ENGINEERING: UX, SEARCH, CHECKOUT

learned early in my career that *how* you sell online is just as important as *what* you sell. You can have the best products and a killer marketing strategy, but if your e-commerce site frustrates people at the UX level – if they can't find products or breeze through checkout – you're leaking revenue. I remember working with a client who was baffled by flat sales. Their ads were driving traffic, but conversions lagged. A quick look at their site told the story: slow-loading pages, a hidden search bar, and a checkout that felt like doing taxes. It was a classic case of ignoring experience engineering. I told them what I'll tell you now: optimizing your site's user experience, on-site search, and checkout isn't a "nice-to-have" – it's make-or-break for your business. These are the moments that decide whether a visitor becomes a customer. Let's talk about why, and how to take action.

First, let's tackle user experience (UX). I'll be blunt – please make your site fast and easy to use, especially on mobile. We've reached the point where mobile shoppers often outnumber desktop shoppers; during a recent holiday season, mobile made up over half of online sales. If your site isn't buttery smooth on a phone, you're turning away half your potential buyers. Speed is a big part of that. Remember, Amazon once found that every 100 milliseconds of page load delay cost them 1% in sales. Think about that: a fraction of a second too slow, and you start bleeding revenue. Users have endless options, and they won't wait – as one analysis put it, even a few seconds of lag can be the difference between getting a sale or losing it to a competitor. Fast sites feel effortless to use, and that creates a positive feedback loop: happy users stick around, browse more, and buy more. So optimize your images, streamline your code, and check your hosting – speed matters because it directly impacts your bottom line.

Of course, UX is more than speed. It's the *whole experience* of navigating and using your site. Is it clear what action you want the user to take on each page? Are your calls-to-action obvious and compelling? Every element should ultimately guide people toward conversion – whether that's signing up for a newsletter or clicking "Buy Now." If there's too much clutter or confusion, people bounce. A good rule of thumb is to design each page with a goal in mind and trim anything that doesn't serve that goal. You want to smoothly lead the user from landing page to checkout, with every click feeling natural. That means intuitive navigation, helpful product filters, and clear CTAs (not a barrage of pop-ups or random banners). And **don't forget mobile.** Chances are the first interaction a customer has with your brand is on a smartphone screen. If your desktop site is beautiful but your mobile site is a hot mess, you've got a problem. I've seen merchants get shocked when they realize an image carousel or fancy layout on desktop turns into an illegible blob on a phone. You have to **design for mobile first**, or at least test everything on mobile as rigorously as desktop. Forms, menus, checkout pages – all need to be mobile-friendly in layout and load time. Most shoppers will *only* experience your store on a mobile device so this isn't optional. Check your site on a variety of phones and tablets; if anything is slow, broken, or hard to use, fix it. Consistency counts too: your brand's look and feel should carry over cleanly to mobile, with easy-to-read text and buttons that don't require precision tapping. The bottom line is a seamless UX makes it easy for customers to say "yes" – and every extra hurdle or annoyance is a chance for them to say "forget it" and leave.

While we're on UX, let's talk inclusivity. A positive experience should extend to *all* users, including those with disabilities. This isn't just about avoiding lawsuits or checking an ADA compliance box – it's a growth opportunity. By making your site accessible (good alt text on images, proper contrast, keyboard navigation, etc.), you open your store to a larger audience that many competitors ignore. People with visual, auditory, motor, or cognitive challenges are shoppers too, and they tend to be loyal to businesses that accommodate them. Accessibility means making your site available to everyone. In other words, better accessibility = more people who can spend money with you. It also often improves your SEO (search engines love clear structure and descriptions) and overall usability for all users. So consider this part of UX hygiene. When you improve your font size, color contrast, or caption your videos, you're not only helping those who need it – you're likely making the experience better for everyone. And that translates to higher engagement and conversion.

Next, let's look at a critical yet often undervalued aspect of experience: **on-site search**. Many merchants set up a search box and forget about it. Big mistake. Think of your search bar as the store associate for your online shop – the person a customer turns to and says, "I need help finding X." If that "associate" (your search function) is ineffective, you're losing sales left and right. Here's a stat worth burning into your memory: **roughly 40% of visitors will go straight to the search bar** when they enter a site, and on-site search users are *2-3 times more likely to convert* than those who just browse. Why? Because people who use search usually *know what they want* – they have intent. They're the closest to making a purchase, if only they can find the item. I've seen this with countless clients: the users who search for a specific product or category are your hottest prospects. But that's only true if your search actually delivers relevant results. If it doesn't? Those high-intent shoppers will bounce to a competitor faster than you can say "no results found." In fact, if customers aren't finding what they expect from your search, or – worst case – your search returns nothing useful, they'll simply give up. It sounds dramatic, but think about your own behavior: how many times will you rephrase a query on a retail site before you just try Amazon or another store? One, maybe two attempts? We don't have patience for poor search anymore.

So *make on-site search a priority*. Ensure your search bar is prominent

– users shouldn't have to hunt for the search function (put it right up top, big and obvious). Implement features like autocomplete and spelling corrections; these guide shoppers and prevent dead-ends. For example, as noted in a UX case study, sites like Swarovski suggest products as you type, which can inspire a click before you even finish typing. That's great UX: it's helping the customer help themselves. Also, analyze your search data. It's a goldmine of insight. What terms are people searching? Are there common phrases yielding zero results? If so, you might be missing products or using different terminology than your customers. Adjust your product names or add synonyms to the search algorithm so that "hoodie" and "sweatshirt" lead to the same results, for instance. One alarming insight I've come across is that only about 15% of companies dedicate resources to optimizing on-site search. No wonder so many get it wrong. Treat search as a core part of the shopping journey: assign someone to regularly review it. When you do, you'll capture more "spearfishers" – those focused shoppers ready to buy – and you'll watch your conversion rate climb as a result.

Now, let's address the final piece of the experience triad: **checkout design**. This is the moment of truth. A customer has added products to the cart – they *want* what you're selling – and now you need to smoothly take their money and give them a feeling of satisfaction and trust. It sounds straightforward, yet the average online cart abandonment rate is around 70%. That means for every 10 people who click "Add to Cart," seven vanish before completing the purchase. Why? Because something in the checkout process spooked them, annoyed them, or otherwise created friction. Checkout *must* be as painless as possible. Every extra step or uncertainty is an invitation for the customer to bail. I cannot overstate this: **simplify, simplify, simplify.** Only ask for information you truly need. If you can, allow things like address auto-completion to save keystrokes. Make error messages clear (and not red text in tiny font that the user misses). And above all, **offer a guest checkout option** – don't force people to create an account just to buy from you. We have a famous example in e-commerce where a company removed the mandatory account registration at checkout and added a simple "Continue as Guest" button. The result? A $300 million boost in sales in one year. You read that right: a single UX change –

essentially one less form to fill out – led to a 45% increase in completed orders. That's how much customers hated being forced to register. They would rather abandon the cart than go through password hell. The lesson: never let your desire for user data or account signups get in the way of a sale. You can always invite them to create an account *after* the purchase (with the carrot of tracking orders or saving info for next time). But during checkout, the motto is "keep it moving." Every click and question you remove counts. If a field isn't absolutely required, ditch it. I often joke (only half-kidding) that the perfect checkout has *two fields*: address and payment. We're not quite there yet for most businesses, but it's a good ideal to strive for.

There are other checkout optimizations that can lift your revenue too. One is being transparent about costs. Unexpected shipping fees or taxes at the last step are conversion killers – they're the #1 reason people cite for abandoning carts. It's far better to show a shipping estimate early or, if possible, bake costs into prices and offer free shipping. Another powerful strategy: offer multiple payment options. Credit/debit cards are standard, but a huge chunk of customers prefer PayPal, Apple Pay, or other digital wallets. Let them pay however they like – the easier it is for them, the more likely they'll complete the purchase. In recent years, **Buy Now, Pay Later** (BNPL) has exploded in popularity for larger purchases. Services like Klarna or Afterpay let customers split payments, and this flexibility often **improves conversion rates and even boosts average order value**. If you sell higher-ticket items or target younger shoppers, BNPL might be a no-brainer; by the holidays of 2024, BNPL accounted for about 9% of online sales and was credited with meaningfully lifting conversions during that season. It's about reducing the psychological friction ("Can I afford this now?") and giving customers options that make them comfortable buying. Lastly, instill trust at checkout. Offer easy access to customer support in case they have a question while checking out. A little reassurance goes a long way when someone is about to enter their credit card details. We want them confident and excited, not second-guessing.

Alright, we've covered a lot of ground – UX, search, checkout – and hopefully you're convinced these are mission-critical areas to constantly improve. Now I want you to do more than just nod along. I want you to **take action**. Knowledge without execution won't move the needle. So here's my challenge: a seven-day quick-win sprint. Seven days, seven specific actions to start engineering a better experience and boosting your revenue. You ready? Let's go:

Day 1: Analytics Recon – Dive into your data. Pull up Google Analytics (or whatever analytics you use) and identify your top 5 most-visited pages. You might be surprised – it's not always the homepage. Also check which pages have high bounce rates or where customers drop off. This will pinpoint where improvements can have the most impact. By the end of today, you should have a short list of pages and funnels to tackle first (for example: "product category X page has tons of traffic but a high exit rate – I need to find out why").

Day 2: Speed Tune-Up – Pick the highest-traffic page from Day 1 and improve its load time. Run it through a tool like Google PageSpeed Insights or GTmetrix. Are there large images that could be compressed? Maybe some script loading that isn't needed? Implement one or two fixes to shave off milliseconds. Remember, even small speed gains can pay off in conversions. For instance, compressing a bloated image banner or enabling browser caching can often be done in a day. Get that page loading faster by tonight.

Day 3: Mobile Usability Audit – Today, experience your site purely as a mobile user. Use your phone (and ask a colleague or friend with a different phone to do the same). Go through key flows: browse products, use the navigation, add to cart, and initiate checkout. Take notes of any frustrating moments – maybe text is too small, a button is hard to tap, or a page layout looks weird on mobile. Fix at least one of those issues today. It could be as simple as enlarging a font, removing a pesky pop-up, or reworking a mobile menu. Ensure that critical actions (like "Buy" or "Search") are front and center on smaller screens. Your goal: make the mobile experience *so smooth* that a first-time visitor could complete a purchase with one thumb.

Day 4: Search Optimization – Time to give some love to your internal search. First, test it yourself: search for 3-5 common products or queries on your site. Do the results make sense? Are top products showing up, or are you seeing oddball results (or none at all)? Next, check your search analytics if available – what are the top queries? Pick one meaningful improvement: maybe you discover people often search "battery" on your electronics store and get no results because your products say "power cell." You could add a synonym or tag to fix that. Or you might realize your search results page lacks filtering options, and you plan a quick tweak to add a filter by category. If you don't have a good search plugin, research one. The aim today is to ensure that a customer using the search bar finds what they're looking for *without frustration*. Given that search users are far more likely to convert, this is low-hanging fruit you can't ignore.

Day 5: Checkout Cleanup – Go through your own checkout process with a fine-tooth comb and ruthless eye. Better yet, watch someone else (friend, family, coworker) try to check out on your site and observe where they hesitate. Today, implement one simplification. For example, if you currently force account creation, enable guest checkout – that alone can be a game-changer if it's not already an option. Or remove an unnecessary field (do you really need a fax number field in 2025?). Maybe you can combine two steps into one, or add a progress indicator so users know it's a short process. If there's a pesky bug (like the "Place Order" button is not obvious on mobile, or credit card errors don't display clearly), get it fixed. Also consider adding a payment method that might be missing – if you don't have PayPal, Apple Pay, or a BNPL option and your audience uses those, set the wheels in motion to integrate one. The goal for Day 5 is to eliminate at least one point of friction in checkout and make the path to purchase that much easier.

Day 6: On-Site SEO Boost – Today is about making sure your site's content and structure help customers *find* you in the first place. Pick one high-value page (perhaps one identified on Day 1 or a product page for a flagship product) and improve its SEO elements. Update the page's title and meta description to be more compelling and keyword-relevant – think about what a customer would search for and make sure those words are present. Ensure the headings on the page are clear and include those key phrases naturally. Maybe add an FAQ section or a bit more detail that answers common customer questions (both enhancing UX and giving search engines more content to chew on). Also check that images on that page have descriptive alt text (which helps accessibility and SEO). These on-site SEO tweaks can increase your visibility for organic traffic, essentially bringing more shoppers to your improved experience. It's a win-win: better content for users and better signals for Google. By the end of today, that page should be more search-friendly than it was this morning.

Day 7: Measure & Plan – One week later, it's time to step back and see what's changed. Check some metrics: did page load time improve on the page you optimized? Any uptick in mobile engagement or a reduction in bounce rate? Test that search query again – now it brings up the intended results. Run through checkout – it's smoother and maybe a few less people are dropping off at the account step (if you removed it). While one week is short to see big shifts in conversion stats, you've laid the groundwork. Document what you did and any early results (even qualitative feedback). Share it with your team if you have one. Most importantly, make a plan to keep the momentum. Perhaps create a simple backlog: list the next five UX improvements or SEO tasks or checkout tweaks you want to tackle in the coming weeks. You might even schedule a recurring "experience review" each month. The point of Day 7 is both to pat yourself on the back for the quick wins you achieved, and to commit to an ongoing process. Experience engineering is never truly finished – it's a cycle of improvements, testing, and learning. But look at what a difference a focused week can make. **Small steps lead to big wins in e-commerce**, and you've just taken several. Great job – keep going!

Seven days, seven wins. None of these alone will transform your business overnight, but together they start to compound. You've plugged some leaky holes in your funnel and set up a smoother ride for your customers. More importantly, you've gotten into the mindset of continuous improvement, which is absolutely vital in e-commerce. I often compare this process to cleaning up that messy garage. You don't tackle it all at once; you take it one shelf, one corner at a time, and before you know it, the whole space is more functional. The same goes for your website: keep chipping away, and you'll build an experience that delights customers and keeps them coming back.

Before we wrap up, let me leave you with this. The changes you made this week – faster pages, a friendlier search, a slick checkout – are not just tweaks, they are revenue drivers. Every improvement is a hurdle removed from the customer's path to purchase. In e-commerce, **ease = profit**. You've seen how treating UX, search, and checkout as critical assets can yield tangible results in conversion and sales. And we're not stopping here. In the next chapter, we'll dive deeper into how you can leverage data without drowning in it – in other words, how to focus on the metrics that matter and avoid being overwhelmed by analysis paralysis. It's time to ensure all these improvements are feeding into smart decision-making. I'm excited to show you how to turn insights into even more action. Until then, keep pushing forward – the experience you create is your competitive edge, so never stop engineering it to be the best it can be. On to the next challenge!

2.3: PERSONALIZATION WITHOUT CREEPINESS

Think about the last time you opened Netflix. It's almost spooky how well it knows your tastes – yet you never feel spied on. Netflix quietly studies what you watch (and what you skip) and uses that data to serve up eerily perfect recommendations. It's so seamless, you just enjoy your binge without a second thought. Now contrast that with a common ecommerce nightmare: you browse for a product – say, a floor lamp – and buy it. Then for weeks afterward, that exact lamp **follows** you around the internet, popping up in ads on every site you visit. I had this happen when we moved into our new office: I purchased a lamp, and for a solid month I saw nothing but ads for the very item I'd already bought. Instead of feeling personalized, it felt like the lamp was stalking me (and as an ecommerce professional, I knew why and I cringed at how much money the brand was wasting on these retargeting ads). One experience is delightfully personal; the other is downright creepy. Which side of that line are you on? Are you delivering the Netflix level of invisible personalization, or are you creeping out your customers with "hey, remember that lamp?" retargeting? It's a gut-check every ecommerce leader should do, because falling on the wrong side of the line isn't just bad for brand vibes – it's likely costing you real money.

Here's the good news: personalization is *still* worth getting right in 2025. In fact, it might be more important than ever. Yes, everyone's buzzing about the latest AI tools and shiny new platform features, but the basics of personalization deliver tangible ROI that's hard to ignore. You can't outspend Amazon – their budget and reach are untouchable – **but you can out-personalize them**. By leveraging your unique brand voice, niche focus, and customer data, you can create experiences Amazon can't. And the payoff is significant: get personalization right and you'll typically boost your average order value by at least **10%**, increase repeat purchase rates by about **15%**, and even see **8% fewer support tickets**. (Yes, better personalization means customers have less need to contact support – because you're anticipating their needs and reducing friction.) Those are meaningful lifts to revenue and efficiency. Remember, it costs about **five times more to acquire a new customer than to retain an existing one** , so nudging those repeat purchases higher is pure gold for your business. In short, personalization done right is a profit center, not a tech buzzword. The key, of course, is doing it *right* – maximizing relevance and value without tripping the creepiness wire.

Over the past few years, working with mid-market brands, I've developed a framework I call the **five-rung personalization ladder**. Think of it as climbing from basic "Hi, $NAME" personalization up to predictive, proactive customer service. Each rung unlocks more value – and requires more sophistication – than the last. Climbing this ladder step by step ensures you build a strong foundation and don't jump into over-engineered personalization before you're ready. Let's run through the five levels:

Level 1: Recognition. This is personalization at its most basic, yet it's amazing how many businesses skip it. Recognition means using what you know about a customer to acknowledge them in a personal way. The classic example: greeting a returning customer by name. Ever gotten an email that starts with *"Hi [Your Name]"* instead of *"Dear Valued Customer"*? That's Recognition. It's a small touch with a big psychological impact – people pay more attention when they feel personally addressed. A quick win here is to **turn on merge tags everywhere** – in your email tool, on your site's welcome message, in SMS –

so that whenever possible, you're saying *"Hi Joshua"* instead of a generic hello. These little tokens of recognition build familiarity. And the hero KPI at this stage is simple: email engagement. Watch your open and click-through rates. If you nail Recognition, you should see open rates climb (more people willing to read an email that greets them by name) and click rates improve as a result. It sounds almost trivial, but I've seen clients get a bump in email opens just by making that one change. If customers aren't even opening your emails, no fancy AI recommendation widget is going to matter. So level one is **know your customer (and let them know you know them)** – it lays the groundwork for everything that follows.

Level 2: Contextual Merchandising. Now we move beyond just using a name, and start using context to personalize what products or content a customer sees. Most ecommerce platforms these days have some built-in capabilities for this – but many brands don't fully utilize them. Contextual merchandising means things like "Customers like you also bought…" or "Complete the look" bundles on product pages. It's showing relevant add-ons or recommendations based on what the customer is looking at *right now*. A classic example: on a fashion site, if you're viewing a dress, the site suggests a matching handbag or shoes (in your size, of course). Or on a electronics site, looking at a camera triggers suggestions for lenses or a tripod. These aren't random upsells – they make sense in context, so the customer might actually appreciate the suggestion. A quick win for level 2 is simply to **enable those native recommendation features** in your platform (many platforms have them). If your platform doesn't have good options, consider a plug-in or third-party service that specializes in product recommendations. The beauty of contextual merchandising is that it directly drives bigger baskets and higher AOV. The KPI hero here is your **upsell/cross-sell conversion rate** – in other words, the percentage of customers who add the recommended item. Keep an eye on that, as well as overall average order value. A healthy upsell conversion rate to aim for might be north of **5%**, and if you're tracking AOV, look for about a **10% lift in AOV** once you implement tailored recommendations. Those are ballpark targets, of course; they can vary from industry to industry. At level 2, you're using *context* (what product or category the customer is engaging with) to personalize in-the-moment and increase relevance – and revenue.

Level 3: Journey Orchestration. This is where we start connecting the dots across the entire customer journey. It's personalization at a higher altitude – not just on one page or one email, but coordinating the experience as the customer moves from channel to channel. If someone browsed certain products on your site, did that data inform the email you sent them later? If they purchased in store, do they get a different follow-up than someone who bought online? Journey orchestration is about **segmenting and syncing** customer data across your e-commerce platform, your email/SMS marketing, your ads, and even your in-store POS (if you have physical stores) to create a cohesive experience. A quick win at this level is to pick one simple segment and build a special journey for it. For example, take new customers (first-time buyers) as one segment and repeat buyers as another. Then orchestrate a post-purchase email flow tailored to each: new customers might get a "Welcome, here's how to get the most from your product" email, while repeats get a "Welcome back, VIP!" treatment with maybe an upsell offer or loyalty program invite. The idea is to **sync your segments and messaging** so that the customer is recognized and guided no matter where they interact. For omnichannel retailers, this absolutely means tying your in-store and online data together – e.g., if I buy something in a store, I should *not* get an email next week pushing me the same item as if I never bought it. Journey orchestration can get complex, but you can start simple. The KPI to watch here is **repeat purchase** rate (within a given time frame). I often track repeat purchase within 60 days as a litmus test for effective journey orchestration. If your campaigns and touchpoints are well-orchestrated, you should see customers coming back and buying again sooner. A good initial target might be to increase your 60-day repeat purchase rate by a couple of percentage points in the first quarter of effort. It shows that your coordinated nudges (emails, ads, recommendations timed right) are shortening the time between purchases. In short, level 3 is where your marketing and site experience start playing in harmony, like a well-conducted orchestra, to encourage loyalty.

Level 4: Predictive AI Bundling. Now we're getting into the fancy stuff. Level 4 uses artificial intelligence and predictive analytics to anticipate what the customer will want next, and bundling or suggesting accordingly. It's like having a really good shop assistant who knows, "Customers who bought X usually need to replenish Y in Z weeks," or "People who bought A and B often love C as well." A practical example is a replenishment program: say you sell a consumable (coffee, pet food, skincare). With predictive personalization, you model each customer's usage and timing, and then you might send them a reminder or even auto-ship a refill right when they're likely running low. In one case study I often cite, a beauty brand set up an AI that would email customers a pre-loaded cart with their likely repurchase items at about the right interval – making it two clicks to re-order via Apple Pay. And this isn't just "every six months" type of logic, it realized, for instance, that certain skincare products were used more often during the summer, and suggested faster replenishment rates based on the local weather. That drove an **18% increase in repeat orders** for them, because it removed friction and hit the timing perfectly. Quick win for you at level 4 could be installing a third-party app that does predictive recommendations (many ecommerce app stores have AI recommendation engines that go beyond the basics of level 2 and approach level 4). If you're not selling something replenishable, predictive bundling could be suggesting a bundle of items that, based on aggregate data, the customer is likely to want together. The key is that it's forward-looking and tailored to *that customer's* behavior, not just generic "people also bought" (which is level 2). The metric to focus on here is the **conversion or opt-in rate for these predictive offers** – for example, what percentage of customers offered a subscription or a predictive bundle actually take it. A healthy benchmark for a well-targeted subscription upsell might be **8–12% opt-in** if you're doing it right. That indicates your AI is predicting well and customers welcome the suggestion. At this level, you're starting to lock in future revenue and build more of a relationship, not just one-off sales.

Level 5: Proactive Service. This is the pinnacle of the personalization ladder – it's where you turn personalization into a form of customer service *that anticipates issues before they happen*. Think of it as the equivalent of a Disney cast member noticing you dropped your ice cream and giving you a new one before you even go ask (Disney actually empowers employees to do stuff like that). In e-commerce, proactive service could mean using data triggers to reach out to a customer before they contact you. For instance, if a shipment is delayed beyond a certain threshold, you automatically send an apology email with a small gift card or promo code for the inconvenience. Or if your system detects that a normally active customer hasn't logged in or purchased in a while, you might proactively check in with them to see if everything's okay or offer assistance. This level often leverages the same data and AI as level 4, but applied to service and support scenarios rather than sales. One simple way to dip your toes in proactive service is to set up an alert for "silent failures" – say, if an order is stuck in processing for more than X days, have someone reach out, or automate an apology email. Customers are blown away when you *notice a problem before they do*. It turns a potential negative review into a positive story. The KPI for proactive service can vary, but I like to use measures of customer satisfaction or dissatisfaction. For example, **negative review rate** (what percent of reviews are 1or 2-stars) or **CSAT scores** in support interactions. If you implement proactive service well, you should see the negative stuff drop. You might aim to cut your negative review rate in half over time, or raise your CSAT by a few points. Another indicator: fewer "Where's my order?" support tickets, since you're solving issues preemptively. At the end of the day, level 5 personalization makes customers feel taken care of **before** they even have to ask – the ultimate trust-builder.

As you climb this ladder, a critical rule is to **tie each personalization effort to a clear metric** (and avoid vanity metrics). Don't just brag about "we send 1 million personalized emails a week" – who cares, if they're not being opened or driving any value? Focus on the outcome, not the volume. To recap the ladder and those key metrics, here's a quick-reference cheat sheet:

Level 1 – Recognition: *KPI:* Email open rate (and click-through rate). *Target:* ~30%+ open rate for your house list (with ~3%+ click rate).

Level 2 – Contextual Merchandising: *KPI:* Upsell / cross-sell conversion rate. *Target:* >5% conversion on recommended products, or about a 10% lift in AOV after adding contextual recommendations.

Level 3 – Journey Orchestration: *KPI:* Repeat purchase rate (within 60 days of first purchase). *Target:* Raise repeat rate by 2–4 percentage points in the first quarter of orchestration efforts.

Level 4 – Predictive AI Bundling: *KPI:* Subscription or auto-reorder opt-in rate. *Target:* 8–12% of eligible customers opt into subscriptions/replenishments when properly targeted.

Level 5 – Proactive Service: *KPI:* Negative review rate or CSAT score. *Goal:* Cut negative reviews roughly in half, or raise customer satisfaction by a few points through proactive outreach.

Now, before you race off to implement all five levels, let's pause and talk about the fine line between personal and creepy. I have a simple **five-question litmus test** I use to ensure we stay on the right side of that line. Anytime you're rolling out a new personalization tactic, ask yourself these five questions:

Would I say this to the customer's face? Imagine walking into a store and saying this personalization out loud to a shopper – would it feel weird? If it would be awkward or invasive in person, it's likely crossing a line online. This gut check helps keep things in the realm of friendly, not stalkerish.

Does it use data the customer knowingly gave me? In other words, am I personalizing based on information the customer *knows* I have? Using their purchase history or profile info is fair game – they expect you to use that. But the moment you start pulling in shadowy third-party data or inferences the customer never explicitly provided, you inch toward creepiness. (There's some nuance here – data enrichment can be powerful, but be very careful with transparency.)

Could this message surprise or unsettle them? You never want a customer thinking, "How on earth did they know that?!" The infamous example here is the Target incident where their predictive analytics allegedly figured out a teenage girl was pregnant before her dad did – and Target sent her maternity coupons, much to the father's shock. True or not, it's a cautionary tale: don't let your personalization reveal something sensitive or premature that might freak people out.

No customer wants to feel like the software knows them better than their family does.

Does it comply with privacy laws? This one's non-negotiable. Between GDPR, CCPA, and others, you must ensure your personalization efforts respect consent and privacy regulations. It's not just about avoiding lawsuits; following privacy best practices also keeps customer trust intact. (And if you're in doubt, consult a legal expert – I'm not a lawyer, and this isn't legal advice!)

Does it add genuine value for the customer? This is *the ultimate question*. Are you personalizing to genuinely help the customer, or just to squeeze more money out of them? Great personalization should feel like a service, not an extractive sales tactic. For example, recommending a camera lens that complements the camera they bought is helpful; bombarding them with unrelated "deals" based on some demographic profiling is not. If it doesn't clearly benefit the customer, rethink it.

If you can answer these five questions in a way that makes you proud of the experience you're crafting, you're likely on solid ground. Essentially, **be the kind of marketer you'd want to encounter as a customer**. By the way, notice how much this litmus test overlaps with basic courtesy and common sense. That's not a coincidence – it leads into another framework I love: Disney's Four Keys.

In case you're not familiar, Disney trains every park employee on the "Four Keys" to a great guest experience: **Safety, Courtesy, Show, and Efficiency**. These were originally about running a theme park (for example, Safety is first – no one should get hurt at Disney; Show means the presentation and magic must be maintained, etc.). But I find the Four Keys make terrific guardrails for personalization strategy too. Here's how I interpret them in our context:

Safety: Protect customer data like treasure. Don't do anything that would put their privacy or security at risk. And it's not just legal safety; it's emotional safety too – not exposing or embarrassing the customer. For instance, if you know a customer's birthday from your loyalty program, using it to wish them happy birthday is safe. Using it to imply you know their age in an ad… probably not safe, because it might feel like an overreach.

Courtesy: This is basically the golden rule. Be polite and respectful with your personalization. Don't shove it in their face that you know things about them; instead, gently *assist* them with that knowledge. Courtesy in personalization might mean giving customers control (like an easy way to adjust their preferences or opt out of certain personalized recommendations) – it shows respect.

Show: In Disney terms, Show is about stage-managing the experience so the customer stays immersed in the magic. For us, "Show" means keep the personalization on-brand and subtle. Everything the customer sees should feel like part of a coherent, delightful experience, not a hodgepodge of recommendation widgets and pop-ups. For example, if your brand voice is fun and quirky, your personalized product suggestions can be phrased in a fun, quirky way (and not come off as robotic). Make the personalized experience feel like it was *made just for me* – that's the goal – without breaking character, so to speak.

Efficiency: Last but not least, make sure your personalization actually makes the experience smoother, not more cumbersome. Personalization should save the customer time or effort (like quickly surfacing products they are likely to want) rather than waste their time. Ever gotten a "recommended for you" that's completely irrelevant? That's inefficiency – it makes the customer do extra work to ignore or close it. Good personalization streamlines the journey; it doesn't slow the user down.

Whenever we brainstorm a new personalization idea, filtering it through the Four Keys helps us see potential red flags. For example, say we want to auto-play a product demo video when a customer lands on a product page because we think it's "personalized content." Is it Safe? Likely, yes. Courteous? Maybe not – some customers might find auto-play video annoying (not very polite). Show? It might disrupt the "story" if it's unexpected or off-brand. Efficient? It could be; a video can quickly show features. That quick Keys analysis might lead us to, for instance, at least make the video opt-in (so it's more courteous). These principles keep us grounded: we want to make each customer feel uniquely valued, but *never* at the expense of safety or trust. When in doubt, err on the side of the Disney magic: make them feel like the hero of the story.

To see how this all comes together, let's look at a few real-world examples. First, consider **Coucou Suzette**, a quirky French DTC brand (they sell fun accessories and jewelry). They decided to move up the personalization ladder by launching a loyalty and referral program called "Club Coucou." Customers could earn points for all sorts of interactions – not just purchases, but things like following the brand on Instagram, sharing content, or writing a review. All that data fed back into how they communicated with each customer. The result? In one year, Coucou Suzette increased their average order value by **54%** and *quintupled* their customer lifetime value (5× LTV). Read that again: 5× LTV in one year. They basically turned personalization into a game for their community, and it paid off. The key was that none of it felt creepy – it was opt-in, it was fun (missions to earn points), and it was clearly about rewarding the customer, not just extracting more money from them. Customers gladly engaged and bought more because the experience was genuinely personalized *and* genuinely delightful. It's a fantastic case of "personalization without creepiness."

Another example: **Allbirds**, the sustainable shoe retailer famous for their wool runners. Allbirds started online (pure-play digital) and later expanded into physical stores, all while maintaining a tight focus on customer experience. Their personalization is subtle but effective. When you're on their website, the product recommendations and messaging you see are heavily informed by your behavior and their overall community data – but it always feels on-brand. If you've bought running shoes, the site might show you new running gear or limited-edition colors of the shoes you like, rather than something random. They also integrate sustainability content (since they know that's a core value for their customers) into the shopping experience. In-store, associates use customer profiles (if you've shared your email) to tailor recommendations – for instance, if you bought something online, they might know not to pitch you the exact same item in store, but rather something complementary. Allbirds' omnichannel approach shows how personalization can span online and offline. It's not intrusive; it's helpful. You feel like *they get you* as a customer who cares about comfort and the environment. While Allbirds hasn't publicly shared numbers that I can cite, their brand's rapid growth and strong repeat business speak to the success of this personal, consistent experience. The lesson: know what your customers value (in Allbirds' case, sustainability and simplicity) and personalize around those themes across every channel.

Finally, a story from a **mid-market brand using AI review mining**. This one comes from a client case study we did. The company had decent sales but an inexplicably mixed bag of customer reviews – some products had a lot of low-star reviews, but the feedback in them was all over the place. We ingested *all* their customer reviews across the product line into an AI tool to find patterns. What we discovered was eye-opening: the primary complaint wasn't about product quality at all – it was about shipping times and delivery expectations. Customers were upset because the website promised, say, 5-day delivery, but items were taking 8-10 days to arrive. This wasn't a product issue, it was a communication issue. Armed with that insight, the company immediately adjusted their shipping estimates to be more accurate (and improved some logistics). The payoff was swift: within a month, they saw a noticeable uptick in positive reviews and a boost in conversion rates, as fewer customers were turned off by the negative comments. This is personalization in the sense of *listening* to your customers and proactively addressing their concerns. By mining review data (which is kind of personalization at scale, learning from individuals' experiences), they improved the experience for everyone. And they did it in a non-creepy way – no individual targeting at all, just using AI to be more customer-centric.

By now you might be wondering, "This all sounds great, but where do I actually start?" Every business is at a different point on the personalization ladder, and the best practices can vary a bit by business model. Here's how I break down the focus areas, depending on what type of business you're running:

If you're a B2B "sidecar" brand (meaning eCommerce is a newer add-on to a traditional B2B business): Start with **Level 2 (Contextual Merchandising)** once you've got the basics of recognition covered. Many B2B sites are behind on personalization, but an easy win is to ensure your catalog pages show related products, accessories, or volume discounts that make sense for your customers. Implement an upsell module that ties into your ERP if possible, so that, for example, when a customer is buying industrial equipment, the site suggests the compatible spare parts or services. Your hero metric here will likely be upsell revenue or cross-sell conversion. Even a modest **5% conversion on recommended add-ons** could translate to significant revenue in B2B, where order sizes are big. B2B buyers appreciate efficiency – if you save them time by highlighting what goes with what, that's value. Just keep it very factual and helpful (B2B folks have an especially low tolerance for gimmicks that feel creepy). Also, leverage your sales reps in this if you have them: MSC Industrial Supply (a $3B distributor), integrated an AI recommendation engine that feeds suggestions to their phone reps during calls. That drove a **20× increase in upsell revenue** because reps could instantly personalize offers for each caller. That's contextual merchandising via the human channel – and it worked brilliantly because it was relevant and timely.

If you're an omnichannel retailer (selling both online and in physical stores): Focus on **Level 3 (Journey Orchestration)**. You want to create a seamless loop between in-store and online experiences. This means integrating your POS with your e-commerce CRM, so you have a single view of the customer. If someone buys in store, your email system should know, and maybe send a follow-up thanking them for the in-store purchase (maybe even with a suggestion for a related item that can be bought online). If they browse online and add to cart but don't buy, maybe a push notification or email could remind them next time they step into your store (geofencing and mobile apps can help with this, if applicable). The KPI to watch here is **repeat purchase rate and frequency**. For example, track how many store purchasers go on to buy online within 60 days or vice versa. Aim to increase that cross-channel repeat behavior; even a few percentage points uptick means your channels are amplifying each other. Another metric: basket size for cross-channel shoppers. One retailer found that customers who used in-store pickup (BOPIS) and received a personalized "while you're here, you might also like…" suggestion ended up with **12% larger baskets** on those pickup orders. That's orchestration at work –

leveraging online data at the store touchpoint to drive more sales (in a helpful way, like, "Grab these related items when you pick up your order"). The key for omnichannel is consistency. Nothing annoys customers more than feeling like two separate people in-store vs online. Personalize in a connected, channel-aware way, and you'll earn their loyalty.

If you're a pure-play digital brand (no physical storefronts, all your sales are online): You should be pushing toward **Level 4 and even 5**. In pure eCommerce, customer expectations are high – they know you have data on them, and they expect you to use it intelligently. If you're not at least dabbling in predictive AI (level 4), you're behind. So for a pure-play, I'd say make sure you've got levels 1–3 running well, then invest in level 4 capabilities like personalized product recommendations powered by AI, predictive replenishment programs, "you might also like" emails based on their browsing history, etc. Set up some proactive service triggers (level 5) as well, because in a digital-only relationship, trust can be fragile – a quick proactive fix to a problem can save a customer from defecting when there's no face-to-face interaction to smooth things over. A good KPI focus here is your **subscription opt-in rate or repeat purchase cadence**. If you sell something consumable, shoot for that **8–12% subscription take-rate**. If not, look at your 1-to-2 purchase conversion: what percentage of first-time buyers come back for a second purchase, and how quickly? Optimize your predictive and journey tactics to boost that. Pure-play brands live and die by LTV and repeat business, so use all the personalization tools to make the experience sticky. One caution: digital shoppers are also the most sensitive to creepiness (since they're used to sophisticated tech). So be very transparent and give control – e.g., if you're recommending products based on what they browsed, maybe a gentle "Recommended for you" label (with an option to hide if they want) is better than pushing it in their face. But done right, advanced personalization is *expected* by online-only customers. If you're not doing it, it's noticeable.

We've covered a lot, so let me leave you with a **seven-day personalization sprint.** In just one week, you can take a meaningful step up the personalization ladder without creeping out your customers. Here's how I'd structure it:

Day 1 – Audit & Pick a Focus: Gather your team for a quick 30-minute "personalization ladder audit." Identify which ladder level is your weakest or most neglected right now. Maybe you realize you haven't updated your on-site recommendations in ages (that's a level 2 gap), or you have no proactive service triggers (level 5 gap). Pick *one* level/rung to improve this week, and decide on a specific micro-test. For example, if Level 2 is lacking, you might decide "Let's enable the 'Related Products' carousel on our top 10 product pages and see what happens." If Level 5 is the focus, maybe "Let's set up a trigger to email customers who haven't received their order in 10 days with an apology and discount." Keep the experiment small and focused.

Day 2 – Assign an Owner & Metric: Clearly assign who is responsible for this experiment. Maybe it's your e-commerce product manager or a specific marketer – someone has to own it. Also, define the success metric for this test (the one KPI that ladder level is tied to). If it's a level 2 test, maybe the metric is upsell conversion rate on those pages. If level 5, maybe it's the number of support tickets about shipping issues. Make sure everyone knows what we're measuring and what success looks like (e.g., "we'd love to see at least a few purchases from the related items carousel this week").

Days 3–6 – Launch the Micro-Test: Implement your experiment and let it run for several days. Resist the urge to tinker constantly – give it a few days to collect data. Monitor it just enough to ensure it's running correctly and not inadvertently creating a bad experience. (For instance, check that the related products being shown make sense and aren't bizarre. Quality control is part of keeping it non-creepy!) Over these days, have the owner gather any qualitative observations too, like customer feedback if available.

Day 7 – Review & Learn: At the end of the week, pull the data on your chosen metric and huddle up the team again. What happened? Did the metric move at all? Sometimes a week is too short to judge definitive results, but you'll often get directional insight. Discuss any anecdotal feedback (did customers click the related items? any comments on the apology emails?) If you saw a positive change –

great! Plan how to roll out this personalization more broadly or optimize it further. If you saw nothing or something negative, that's fine too – it's a cheap lesson. Maybe the execution needs tweaking, or maybe that rung isn't where you'll get the best ROI right now. The point is to **learn** quickly. End the meeting by deciding on a next step: scale it, tweak it, or try a different ladder rung next sprint. And celebrate that you ran a focused experiment; too many teams plan massive personalization overhauls and never get past planning. You *did* something in a week, and that momentum is gold.

Rinse and repeat this cycle, and over a few months you'll likely climb at least one full rung on the personalization ladder. Each micro-test builds your personalization muscle, and because you're measuring everything, you'll have proof of what's working (and you'll avoid big, costly, creepy missteps). Keep championing the idea that personalization is a process, not a one-time project. It's like fitness – consistent small workouts beat a once-a-year extreme crash diet.

I'll close with this: personalization is ultimately about **respect and relevance**. When you get it right, customers love it – they feel like *you just "get" them*. When you get it wrong, they feel icky. But now you've got the tools, framework, and examples to get it right. Start small, stay honest (use the litmus test), and build on each success. Your customers will reward you with loyalty and bigger purchases, and you'll sleep well at night knowing you're delivering personalization with integrity.

And if you're excited about this, just wait – next up we're diving into the world of **Omnichannel**. That's the chapter where we'll connect all these personalized experiences across every customer touchpoint. From online to in-store to social and beyond, we'll ensure your brand delivers a cohesive, channel-agnostic journey. So, take a breather, maybe implement a quick win from this chapter, and get ready: in the next chapter, we'll tackle how to bring your personalization prowess into the omnichannel arena, keeping the magic alive everywhere your customer goes.

2.4: OMNICHANNEL THAT ACTUALLY WORKS

n 2020, Best Buy's omnichannel strategy helped it **surge 242% in online sales** despite pandemic store closures. Curbside pickup and turning stores into mini-distribution centers kept customers shopping and "canceled out headwinds" from lockdowns. Meanwhile, Nordstrom Rack tried to ride the same wave – and wiped out. The off-price chain's **inventory inaccuracies and clunky store fulfillment** led to so many canceled pickup orders that Nordstrom **pulled the plug on Rack's BOPIS program** altogether. Digital sales at Nordstrom Rack **fell 16%** after it scaled back omnichannel, a stinging reminder that if you get this wrong, customers (and sales) walk away.

Why the stark contrast? Omnichannel isn't a buzzword – it's the core margin battleground of retail in 2025. Best Buy won because it executed omnichannel with precision and speed; Nordstrom Rack faltered because a half-baked approach chewed up profits. In this chapter, I'll show you how to make omnichannel actually work – in practice, on the P&L, and for your customers. We'll break down the three pillars of omnichannel success, five real operating models you can pilot in 90 days (with concrete KPIs to prove impact), and a few cautionary tales so you don't repeat costly mistakes. I'll also share quick-start playbooks for different business types – from a side-car B2B venture to legacy retailers and pure-play D2C brands – and map out the tech stack ladder that enables these models. In true playbook fashion, we'll close with a 7-day quick-win sprint to build momentum, and a sneak peek at what's next. Let's dive in.

OMNICHANNEL: THE MARGIN BATTLEGROUND OF 2025+

Omnichannel has become **ground-zero for retail margins**. Consumer expectations for free, fast delivery and seamless store pickup are higher than ever, putting pressure on retailers to deliver convenience *without* eroding profits. Consider that most shoppers now expect 2-day shipping at no extra cost, yet the *actual* cost of fulfilling omnichannel orders can run **10–20% of sales** – a huge bite out of margin. If you're not careful, offering "buy online, pick up in store" (BOPIS) or free ship-to-store can turn into a margin sink.

Done right, though, omnichannel can **boost your profitability and customer spend**. A study of 46,000 shoppers found that the more channels customers use, the more they buy – **omnichannel shoppers spent 9% more in-store** on average and logged 23% more repeat trips than single-channel shoppers. The key phrase is *done right*. A recent analysis showed that **same-day store pickup can *lower* margins by up to 6.8 percentage points** for retailers whose systems and processes aren't optimized – essentially a profit leak. But retailers who **build efficient omnichannel ops actually see a margin *lift* of ~3.2 points** on pickup orders. In other words, **omnichannel is a make-or-break efficiency play**. It will either streamline your cost structure and drive higher lifetime value, or it will introduce expensive chaos – depending on how you execute.

This is why I call omnichannel a *battleground*: in 2025, it's where your retail business wins or loses on both the top and bottom line. Every CEO's earnings call now highlights omnichannel performance, and investors know that **blending physical and digital channels is essential for sustainable growth**. Customers don't think in "channels" – they just shop. It's up to us to break internal silos and serve that seamless experience *profitably*. The good news: we have a blueprint. It comes down to getting the basics right – *Inventory Truth, Fulfillment Flexibility, and Unified Data* – and then building on that foundation with agile experimentation. Let's break down these three pillars of omnichannel success.

The Three Pillars of Omnichannel Success

Pillar 1: Inventory Truth. This is the non-negotiable foundation: a **single, real-time version of inventory across all channels**. If your website thinks an item is in stock but the store shelf is empty, you've set up a customer for disappointment – and likely lost the sale. I often say *"you can't sell what you can't find,"* and it's painfully true. Industry research shows store-level inventory records can be *shockingly* inaccurate – **as low as 60% accuracy in some cases**. Think about that: nearly half of the products might be misplaced or miscounted. In today's world, that is **catastrophic for omnichannel**. Nordstrom Rack learned this the hard way when their BOPIS rollout imploded – store associates simply couldn't locate items that the system said were there, leading to **abysmal order cancellation rates**. If *Nordstrom* can't get inventory right, it begs the question: how many retailers are flying blind on what's *actually* on the shelves?

Achieving **"inventory truth" means real-time (or near real-time) inventory updates, robust auditing, and zero tolerance for phantom stock**. In my own projects, I've made this pillar a first priority. For example, when we built omnichannel for a 150-store farm & home chain, we had **over 100,000 SKUs per store** and absolutely could not afford to sell something online that had just sold out in-store. Our solution was a 15-minute batch inventory update, syncing millions of SKU-store combinations every day to the e-commerce site. It was a Herculean integration with their ancient Epicor Eagle ERP (flat files and all), but it worked – **inventory data updated every 15 minutes** meant customers could trust that "available for pickup" actually meant *available*. That's the level of commitment needed. If your system can't handle that, implement a process that can – even if it's scrappy at first (e.g. hourly CSV inventory feeds, manual adjustments, etc.). The payoff is huge: **near-perfect inventory visibility is what powers BOPIS and ship-from-store reliability**. Retailers using RFID and better tools are pushing 98-99% inventory accuracy, and it shows in customer satisfaction. *Bottom line: Without inventory truth, none of your* *omnichannel promises are reliable.* Invest in the people, process, and technology to know your stock – everywhere, all the time.

Pillar 2: Fulfillment Flexibility. Omnichannel isn't one-size-fits-all; it's about giving customers **multiple ways to get their stuff** – and being able to pivot when conditions change. The more *fulfillment flexibility* you have, the more resilient and responsive your business will be. Best Buy's success during COVID is a case in point: they rapidly shifted to curbside pickup and **appointment-based shopping,** essentially *overnight,* because they had the pieces in place (inventory visibility, an order pickup system, and trained staff) to offer different fulfillment options. Customers loved it. In contrast, many retailers without these options were dead in the water until they scrambled to catch up.

Fulfillment flexibility means **offering and excelling at:** BOPIS (in-store pickup), curbside pickup, ship-from-store, ship-to-store, and even vendor-direct fulfillments. You might not deploy all of these at once, but you need the capability to mix and match. For example, early in the pandemic I worked with Family Farm & Home (a regional retailer) to stand up an **"Insta-BOPIS" curbside program in literally one week** – we limited it to crucial SKUs like livestock feed and medicine for immunocompromised customers, and just **hustled a bare-bones curbside system out the door**. Was it elegant? Nope – customers would place orders online for select items, we'd send an email when ready, and they'd call the store when parked out front. But it met the moment: we likely saved lives and certainly won loyalty by being *that fast* and flexible. Speed beats perfection in omnichannel. You can always optimize later (we eventually integrated SMS alerts and a nicer scheduling interface), but **be ready to flex** when customers or conditions demand it.

Another aspect of flexibility is the **"endless aisle" capability – fulfilling orders beyond your store's four walls.** A great example is how we helped Rural King implement **vendor ship-to-store** for special orders. Before, if a customer in the store wanted an out-of-stock tractor part, store clerks had a dusty special order pad that was a black hole. We digitized it so that the **vendor's warehouse would ship the item to the customer's local store** (often hitching a ride on the regular truck to that store), and the customer got a pickup notification like any normal BOPIS order. Suddenly, a paper process became a seamless part of the online catalog – we expanded selection without carrying more inventory, and customers appreciated the convenience. Fulfillment flexibility also means **using stores as shipping hubs** when it makes sense. Target is the poster child here – its stores fulfill *over 95%* of all orders, leveraging their proximity to customers for faster delivery at lower cost. We'll discuss ship-from-store more in a moment, but the takeaway is: **make your fulfillment network adaptable**. Warehouse, store floor, curbside, supplier – it shouldn't matter to the customer, and you should be able to route orders optimally based on who can serve the customer fastest and cheapest. If one channel gets disrupted (say, a DC delay or a sudden store closure), a flexible operation can shift load to another channel with minimal friction.

Pillar 3: Unified Data & Identity. The third pillar is about **connecting the dots on customer data and engagement across channels**. In an omnichannel world, a customer might browse your website, click an Instagram ad, visit your store to see the product, then buy it later via your app for curbside pickup. If each of those touchpoints is siloed – different systems, no shared customer profile – you're flying blind and likely delivering a fragmented experience. Unified data and identity means having a **single view of the customer and product data** that drives every interaction. Practically, this looks like unified customer accounts/loyalty (so that an in-store purchase links to their online profile), centralized product and pricing data (so that an item has one description and price whether on your site or point-of-sale), and integrated analytics to track the *whole* journey.

Why is this so critical? Because **73% of customers use multiple channels during their purchase journey**. Most of your shoppers are omnichannel shoppers already. If you're not unifying their data, you can't understand or serve them well. For instance, if a customer frequently orders online but also routinely comes in-store, your marketing should reflect that – maybe emailing them about an in-store event or giving them an "online order, pickup in 30 minutes" guarantee because you know they like quick pickups. These kinds of personalized touches are only possible when **your systems talk to each**

other and present a cohesive customer identity. It's not just about marketing; unified data helps store associates too. I've seen clients empower their store staff with clienteling apps that show what a customer browsed online or their past purchases, so the associate can give relevant advice on the spot. On the back end, unified data means **consistent KPIs** and decision-making. You're not stuck debating "online vs store" sales because you can see the interplay (e.g. how online research drives store purchases, etc.). We'll get to metrics soon, but know that a **single source of truth for customer, inventory, and order data** is the heartbeat of omnichannel.

On a practical level, this pillar often involves investing in a **Customer Data Platform or integrating your CRM/POS/e-commerce data** to create that single customer view. It might mean linking your loyalty program across channels (so points from store purchases show up in the app, for example). It definitely means killing duplicate data silos – your "online customer" and "in-store customer" should be the *same customer*. One of my favorite small-scale examples is a 3-store chocolate shop I worked with. They used Square for POS and integrated it into their Adobe Commerce website, which meant from day one, **their in-store and online sales shared one payment system and customer database**. With just 50 SKUs, they pulled off seamless in-store pickup and curbside with almost no budget, because the data and tools were unified by default. That's a win for unified data – even a micro-chain can do it with the right platform choice. If they can do it, you can too. The big takeaway: **treat omnichannel as "one channel" on the back-end**. Everyone in your org should be looking at the same customer profiles, the same inventory counts, and the same order queue, whether the order came from the web or the store. That unity enables all the cool omnichannel experiences *and* gives you the insight to keep improving them.

With these three pillars – **Inventory Truth, Fulfillment Flexibility, and Unified Data** – in place, you've got the infrastructure for omnichannel success. Now it's time to talk execution. How do you actually roll out omnichannel capabilities without spending three years in planning? Let's look at **Five Operating Models** you can deploy quickly, each with a 90-day MVP plan and real-world KPIs to measure.

These are road-tested plays I've run with clients, and you can mix and match them to suit your strategy.

FIVE OMNICHANNEL OPERATING MODELS (WITH 90-DAY MVPS & KPIS)

Omnichannel isn't a monolith; it's a set of tactics and models you can implement. Here are five core models, each a building block of an omnichannel strategy. You don't have to do all five at once – pick the ones that address your biggest customer needs or operational gaps. For each, I'll outline a **90-day MVP (Minimum Viable Pilot)** – what you could get up and running in roughly a quarter – and the **key KPIs** to prove it's working.

Buy Online, Pick Up In-Store (BOPIS): This is the starting point for most brick-and-mortar retailers going omnichannel – BOPIS is table stakes. It's what people expect. If you have physical stores, offering in-store pickup for online orders is almost *mandatory* now. Customers love the convenience (no shipping fees, get it same-day) and it drives foot traffic to your stores. **90-Day MVP:** Don't over-engineer it. Pick one or two pilot stores and enable a basic BOPIS flow for a subset of products. For example, enable store-pickup for your top 100 SKUs that you're confident in inventory-wise. Use whatever tech you have – even if your e-com platform lacks fancy BOPIS features, you can hack it by letting customers choose "Store Pickup" as a shipping option and sending the order info to that store's email or printer. Set up a small pickup area or shelf, and train a couple of staff on a simple process: receive online order notification, pick and set aside the items (maybe with a big post-it or printed receipt attached), and notify the customer (manually if needed). **Key KPIs:** Track the *number of BOPIS orders* per week and the *pickup time* (time from order to ready-for-pickup notification). Also measure *customer uptake* – are customers actually selecting the option? Even a handful of orders is a win initially. Watch for any *stock-outs or substitutions*, as these indicate inventory accuracy issues to fix. One real-world example: our small chocolatier client saw BOPIS orders almost immediately once they enabled it – locals loved reserving online and knowing their treats would be waiting in-store. We measured that online conversions improved because shipping was no longer a barrier for nearby customers. **Pro tip:** Invest in signage early. Clear signs in-store (and curbside) for "Online Order Pickup" will make a world of difference – customers shouldn't wander around confused. Consistent signage across all locations was one of my *exec must-haves* on a project, because it both guides customers and signals to your staff that this is a serious, permanent service.

Curbside Pickup: Closely related to BOPIS, curbside is essentially BOPIS taken to the parking lot – and it's become a customer favorite for convenience. It's worth calling out separately because the operational tweaks are unique. Customers who choose curbside want **speed and zero hassle** – they literally don't want to get out of the car. **90-Day MVP:** If you already have BOPIS, curbside is an incremental add-on. Designate a couple of curbside parking spots (get signs made or even temporary cones), and set up a way for arriving customers to notify you. The MVP version? A cheap prepaid cellphone at the service desk that customers can call or text when they arrive (Include this number in your pickup confirmation email). Or use a Google Voice number that texts store staff. Train your staff to bring orders out within e.g. 5 minutes of the call. Equip them with whatever they need – maybe a rolling cart, and definitely make sure they have the order details and the customer's car description. During our *one-week curbside blitz* for Family Farm & Home, we used simple colored tags: customers told us the color/make of their car when calling, and we matched it with a tag on their bag. Low tech, but it worked for the initial surge. **Key KPIs:** *Customer wait time* (from "I'm here" alert to item in trunk) is critical – aim for under 5 minutes as a starting goal. Also track *participation*: what % of online customers opt for curbside when offered? And *completion rate*: did any curbside orders end up going uncollected (which could indicate communication issues)? During the pandemic, curbside went from 0 to significant double-digit percentages of orders for many retailers – and a lot of folks never went back to walking in. Even post-pandemic, curbside remains popular for grocery, home improvement, and other sectors (because frankly, we're all busy and a bit lazy – don't underestimate the laziness of the American consumer). One more KPI to watch is *upsell or attachment rate*: even though curbside customers aren't browsing in-store, you can still drive impulse buys (e.g. promote an add-on item in the pickup confirmation or have a flyer/sample to give at curbside). Measure if curbside customers are responding to those cues.

Ship-to-Store (Site-to-Store): This model allows customers to buy something online and have it delivered to your store for pickup, often for free. It's a great option if an item is not stocked at that local store or if the customer wants to save on shipping fees. It also has the benefit of drawing the customer into the store to pick up, where they might buy more. **90-Day MVP:** To pilot ship-to-store, you can start by enabling it for items that are *only available online*(e.g. extended sizes, web-only SKUs) as a value-add service. Technically, if your e-commerce can ship to any address, you can make your store address a "shipping" option for the customer. MVP hack: Add each store location as a selectable shipping address in the checkout dropdown (or simply instruct customers: input Store #123 address for free delivery). A bit clunky, but it works to test demand. Simultaneously, prep your store staff: they'll receive these packages (likely via your normal DC-to-store truck or UPS) with the customer's name, and they need a procedure to log it and notify the customer. Use your existing notification for BOPIS – just treat these like any pickup order. **Key KPIs:** Track *how many customers choose ship-to-store* when offered, and the *lead time* (order placed to ready at store). Also measure if those customers show up promptly and whether they purchase anything else during pickup. One real metric I've seen: when we enabled ship-to-store for a fashion retailer, about 20% of customers who lived near a store chose the free ship-to-store option over paying for home delivery. It cut our shipping costs (multiple orders would batch to stores in one shipment) and about 30% of those customers bought an extra item when they came in to get it – either an accessory or another on-sale item. That attach rate was a key KPI demonstrating the store visit value. **Note:** Be mindful of *thresholds* – one client I work with recently raised their minimum order for free ship-to-store because they found small orders weren't profitable. So as you pilot, note if people are abusing it for say a $5 item. Set sensible policies (maybe "free ship-to-store for orders $25+") if needed.

Ship-from-Store: This is a more advanced but high-impact model: using your store locations as fulfillment centers to ship online orders to customers. It can significantly speed up delivery (sending from a store just 20 miles from the customer instead of a DC 5 states away) and can improve sell-through (stores can ship out excess inventory instead of marking it down). **However,** it's operationally complex – this is where many omnichannel efforts get bogged down or have challenges building out the business case and processes. **90-Day MVP:** Start *small*. Identify one or two stores (or a region) where inventory is healthy and staff can handle extra tasks. Also choose a subset of SKUs that make sense – ideally small/medium items that are easy to pack, and perhaps high-margin or slow-moving products that you'd love to sell at full price to a broader online audience. For example, one outdoor gear retailer I know piloted ship-from-store but **excluded bulky items like kayaks** at first – they focused on apparel and accessories that their stores could pack easily. In 90 days, you can integrate a lightweight process: use your e-commerce system to reroute eligible orders to the pilot store (even if it's manual – e.g. an email alert to the store for any web order in their region). Equip the store with basic packing materials (boxes, tape, a scale, printer for labels). You might not have a full Order Management System routing logic yet – that's okay, start with something like "if customer zip is within 100 miles of pilot store, we'll fulfill from store." Monitor results and refine rules as you go. **Key KPIs:** The big ones: *Order fulfillment time* (can the store ship as fast as the DC?), *shipping cost* (are you saving last-mile costs?), and *inventory turn* (are stores reducing overstock?). Also watch *order cancellation rate* closely. A high cancellation or split-shipment rate means the store inventory wasn't as reliable as you thought (sold something they didn't actually have on hand – fix your counts before scaling). Another KPI is the *impact on store ops*: measure how many orders a store can ship per day without hurting in-store service. During a pilot, we had one store manager who deprioritized online orders (they sat in an "exception" queue as overdue orders) – and another manager who embraced it and hustled, shipping same-day. The latter saw **measurable revenue lift** in her store's numbers from fulfilling online demand. Sharing those wins (and perhaps providing incentives for stores) is key to scaling ship-from-store. One caution: **packaging and logistics costs.** It's not glamorous, but you need to figure out boxes and packing materials. I had a client point out -- if we need 5 box sizes across 500 stores, that's tens of millions of dollars in cardboard! So track the *packing cost per order* as you pilot; you might find you need to standardize a couple of box types or supply more materials as volume grows. Overall, ship-from-store can be a huge for omnichannel, but approach it methodically – prove it out in one area, get the kinks worked out (especially inventory accuracy and staff training), then expand.

Vendor Direct-to-Store (Endless Aisle): We touched on this earlier – allowing customers to purchase items that *you don't stock*, by partnering with vendors to ship directly to your stores or to the customer. There are two flavors: **ship to store for pickup** (vendor sends item to your store, customer picks up there), or **drop-ship to customer** (vendor ships straight to customer on your behalf). Both expand your assortment without tying up capital in inventory, but require good coordination. Let's focus on the ship-to-store variant, since it keeps the experience in your stores. **90-Day MVP:** Identify a few high-demand products or categories that you often *"could have sold"* if you had them. Typically these are either oversized items you can't stock everywhere, or long-tail SKUs, or maybe new product lines you want to test. Approach the vendor of those products with a simple proposal: you'll list their item on your website (or in-store kiosk), and when an order comes in, the vendor will ship it to the designated store. Outline an SLA (e.g. must arrive in 5 days), and decide who pays shipping (often you can negotiate free or low cost by bundling it with regular orders from the vendor). In the MVP, you can even handle orders manually: send the vendor an email or place a wholesale order yourself for that item to the store's address when a customer buys it. It's low-tech, but proves the concept. **Key KPIs:** Track *incremental sales* – items sold via this endless aisle that you otherwise would not have sold. That's pure upside (aside from any drop-ship fees). Also monitor *customer satisfaction*: did these special orders arrive on time at the store? Were customers happy with the pickup experience? Since this involves another party (the vendor), measure their *on-time delivery rate* to your store. In our Rural King project, we automated vendor-ship-to-store via EDI and APIs after the MVP phase, and it became a substantial revenue stream – effectively turned hundreds of stores into pickup points for an expanded catalog. One thing we learned was to **negotiate**

vendor agreements upfront: have clear rules on who updates the customer, who handles returns, and maybe even penalties for late delivery. For your pilot, keep it to one or two reliable vendors you trust. If it works, you can scale it into a formal endless aisle program. The KPI to watch long-term is *attachment rate* here too – customers coming to pick up a special order might browse and buy other stuff in-store, which makes your stores more relevant. This model essentially blends e-commerce endless assortment with the convenience of local pickup and the personal touch of your store team.

Implementing any of these five models will push your organization to break silos and iterate fast. A crucial lesson I've learned: **get cross-functional buy-in early**. Omnichannel touches e-commerce, store ops, inventory planning, customer service, you name it. If one team isn't on board (or worse, feels threatened), you'll encounter turf wars and friction. For example, I've seen situations where store managers felt like fulfilling online orders was "not my job" and deprioritized it – until we showed them the sales uptick and gave credit for those sales to their P&L. Suddenly, skeptics became champions. So as you roll out these models, bring everyone to the table: retrain store staff, adjust compensation if needed, and celebrate the wins jointly.

Before we move on, let's step into the **caution-tape zone** and examine a few omnichannel failures (and near-failures). I firmly believe you can learn as much from what *not* to do.

Omnichannel Failures & What Went Wrong

The Nordstrom Rack Meltdown. This fiasco has already been mentioned, but it's worth dissecting. Nordstrom Rack's attempt at BOPIS and ship-from-store failed so badly that they publicly shut it down in 2022. Why? In short: *terrible inventory accuracy and a misfit model*. Rack stores are "treasure hunt" environments – lots of unique clearance items, constantly shifting. Their systems showed inventory that **associates simply couldn't find on the floor**. High cancellation rates ensued, frustrating customers and employees alike. It got so bad that leadership admitted it was **"more difficult finding the product in a treasure hunt environment"** and that the economics didn't pan out for low-price items. The cautionary tale: **Don't force an omnichannel model that your operations can't support.** If your stores run like chaotic clearance bins, either invest heavily in inventory management first or reconsider offering store pickup at all. Nordstrom Rack also underestimated the cost side – fulfilling a $15 item from a messy store ate up any profit, especially when half the orders couldn't be fulfilled on first try. This failure highlighted the importance of Pillar 1 (Inventory Truth) in glaring neon. As one analyst quipped, *"If Rack's inventory is so inaccurate they can't do ship-from-store, how much of their **financial reporting** is inaccurate?"*

Operational Snafus and Internal Turf Wars. Technology isn't the only failure point. Often, *people and process* break down. One common pitfall is not training store staff or allocating resources for new omnichannel duties. I've seen retailers launch BOPIS, but fail to designate staff to pick the orders. Customers show up at the pickup counter and nobody knows what's going on – the orders are sitting in a printer tray or an email inbox unnoticed. That's a fast lane to angry customers. The fix is obvious: assign clear responsibility and ownership in-store, and provide proper training (and incentives). In one case, we set up a simple **red-yellow-green light system at the store's pickup counter** – a physical tower light that turned green when an order was placed, yellow as it neared pickup SLA time, and red if overdue. It was literally a beacon that said "hey, there's an online order to handle!" – which solved the problem of store staff ignoring the digital queue. Use whatever tricks necessary to integrate new workflows into the bustle of store operations.

Another caution: **internal conflict between e-commerce and store teams**. Omnichannel blurs the lines, and if your organizational structure or incentives pit channels against each other, you're asking for trouble. I've seen a scenario in a B2B context where the direct web sales team and the field sales reps basically went to war over orders – the reps felt the website was stealing their commissions. It got comical, like an episode of *The Office*, but it was a real business issue. The company had to implement clear rules and compensation adjustments so that online sales in a rep's territory still benefited that rep, ensuring everyone was aligned. In retail, this can happen if store managers don't get any credit for online orders they fulfill or if online gets all the glory for sales that stores assist. **Align incentives and metrics** so that *shared success* is the goal. The exec mantra should be "one team, one goal" – perhaps using an omnichannel sales metric (online + a percentage of store pickup sales) as part of evaluations.

Logistics and Cost Overruns. Be wary of underestimating the nitty-gritty details – packaging, shipping costs, software fees – that can turn a promising omnichannel idea into a money pit. The story about "millions in cardboard" cost for ship-from-store was no joke. A large-scale rollout means supplying every store with packing materials, printers, maybe even refrigeration units (for grocery curbside) – it adds up fast. Retailers like Target and Walmart have poured capital into store retrofits to enable these services (dedicated pickup areas, cold storage for groceries, etc.). You might not need all that at first, but **do the math before scaling**. Run a simple model: if each store needs $X of gear and Y hours of labor daily for omnichannel orders, at what order volume do you break even or start losing money? Ensure your pilot data informs this. Studies have shown that if your omnichannel processes aren't optimized, you could lose ~6-7 margin points on store pickup orders. IHL Group found that **retailers well-equipped for BOPIS** *gain* **margin, those who aren't** *lose*. So don't let hidden costs creep in unmanaged – optimize processes, negotiate better shipping rates for store-originated parcels, maybe use curbside to cut carrier costs, etc.

Finally, a classic failure zone: **technology overload or mis-selection**. Some companies tackle omnichannel by immediately buying an expensive "omnichannel platform" or OMS, assuming it will magically solve their problems. I can't stress this enough: **tech is an enabler, not a silver bullet**. I've built custom omnichannel solutions on relatively simple tech stacks that outperformed big vendor suites, because we tailored it to the real-world needs and kept it user-friendly for store associates. Conversely, I've seen retailers drop seven figures on an out-of-the-box OMS only to shelve it because it was too rigid or complex for their team to use. You might not even need to buy any other technology; you just need someone to put your existing technology together the right way. Focus on **process and people first**. Get a basic solution working (even if it's driven by scripts, cron jobs and elbow grease). Once you prove the value, *then* consider scaling up to more automation or a more robust system – ideally when you have a clear list of requirements that the fancy software must meet (and you've vetted that it actually can). In short: **avoid "shiny object syndrome."** The latest AI-driven omnichannel predictive blockchain whatever won't save you if your fundamentals are broken. Nail the basics, then smart tech can accelerate you.

By now you should have a realistic sense of the challenges and how to avoid the big pitfalls. Let's switch gears to a more optimistic lens: how different types of businesses can *win* with omnichannel. Not every company will use the same playbook, so I've sketched out three archetypes with quick-start strategies for each.

Omnichannel Archetypes & Quick-Start Playbooks

Every business is unique, but many fall into a few broad archetypes when it comes to omnichannel readiness and needs. I'll cover three here – **Side-Car B2B**, **The Omnichannel Retailer**, and **Pure-Play Digital** – and outline how each can quickly leverage omnichannel tactics for growth. Think of these as tailored mini-playbooks.

Archetype 1: Side-Car B2B

This is a scenario I'm seeing more often: a company that traditionally sells B2C (direct to consumers) wants to add or scale up a **B2B component** – selling to businesses, organizations, or bulk buyers – *without* disrupting the core business. It's like attaching a side-car to your motorcycle; you're adding a new passenger (revenue stream) while the main bike keeps running. For example, a home goods e-tailer might realize small offices want to buy their furniture in bulk, or a farm supply retailer discovers local landscaping businesses buying tons of seed and feed. **Quick-Start Moves:** The first step is to **identify the specific needs of your B2B customers**. Do they need bulk pricing? Tax-exempt purchasing? Purchase orders and terms? Often a side-car B2B can start by simply tweaking your existing e-commerce site: enable account-based pricing (even a basic tier like "login for business pricing" plugin), or create a separate portal page for business customers with a contact form or order spreadsheet upload. In 90 days, you can launch a bare-bones **B2B ordering portal or process**. I helped a client set up a "business desk" email – essentially a concierge service for B2B orders that didn't fit the consumer checkout. Not fancy, but it allowed them to start capturing those sales immediately while we worked on a more integrated solution. **Key Considerations:** Watch out for channel conflict – if you have a sales rep team or distributors, assure them this is about expanding the pie, not undercutting them. Align pricing so your website isn't accidentally cheaper than your wholesale rates, for instance. In one case, an internal sales team felt threatened by the new B2B web portal – we addressed this by assigning inbound web leads to reps and giving credit, plus ensuring pricing parity. Communication and transparency are critical here. **Omnichannel Tie-In:** Even though B2B buyers are businesses, they're still humans who have come to expect consumer-like convenience. I often tell clients, *B2B buyers demand that same seamless experience* – they want to see stock levels, get quick delivery, and maybe pick up at a store if urgent. So leverage your omnichannel capabilities for them too. For example, allow a business buyer to order online and pick up in a store (perhaps with a larger order staging area). Or if you have both B2C and B2B, unify the inventory pool so a product sold to a business client online also reflects in store availability. A quick win for side-car B2B could be listing some of your products on **marketplaces like Amazon Business** to reach more buyers – "go where the customers are" is especially true in B2B. Just be ready: those channels have strict data and shipping requirements. In summary, a side-car B2B strategy in omnichannel means *adapting your current assets to a new audience.* Start with manual or low-tech solutions to validate demand, then iterate. One of my upcoming projects is a 4,000-word guide on modern B2B e-commerce success – trust me, it's a deep topic – but the core advice is: **start simple, involve your sales teams, and deliver consumer-grade service to your business buyers**.

ARCHETYPE 2: THE OMNICHANNEL RETAILER

This archetype is the most obvious: you have physical stores and an online presence – now you need to marry them into a cohesive omnichannel machine. Most mid-market retailers fall here. The good news is you have the infrastructure; the challenge is integration and execution. **Quick-Start Moves:** If you're just beginning, implement BOPIS *yesterday*. Seriously – if you still don't have buy online, pick up in store, that's priority one. As we established, it's expected by customers, and it's likely already happening informally (customers call stores asking to hold items they saw online, etc.). Next, focus on **inventory integration**: get your POS and e-commerce inventory talking in real time (even if via flat-file updates like we did for Rural King due to their aging ERP). A unified inventory feed is the backbone for all store-related fulfillment. Another key move is to **train and incentivize store staff** for omnichannel tasks. Make it part of the culture that an online order is as important as an in-person customer. One client of mine made it fun by running *internal competitions*: which store can fulfill online orders fastest, or who gets the highest pickup customer satisfaction scores. Winning stores got recognition and small bonuses. It turned a chore into a game and boosted performance. **Leveraging Stores:** As an omnichannel retailer, think of your stores as assets for more than just sales floor transactions. They are mini-warehouses (for ship-from-store, as discussed), they are pickup hubs, they can even be service centers (handling online returns, etc.). One playbook item: enable **cross-channel returns** – let customers buy online and return to any store. It's a bit of extra work to handle those returns, but it removes friction and draws that customer into the store (where they might exchange or shop more). We measured this at a retailer: allowing in-store returns for online purchases not only improved customer satisfaction but about 10% of the time the customer bought something else during the return visit, salvaging the sale. **Data Unity:** Ensure that your loyalty program or promotions work across channels. If a customer gets a coupon via email for online use, let them scan it in store too. Conversely, if they earn loyalty points in store, those should show up in their online profile. These things sound obvious, but I've seen many retailers struggle with fragmented systems where store and online loyalty were separate – leading to customer confusion and irritation. A quick integration (or even a nightly sync of loyalty points between systems) can solve that. **Example – Rural King:** I'll use our Rural King project as a mini-case. They embraced being an omnichannel retailer by doing some heavy lifting: upgrading systems, building a custom headless **order management system** to fill gaps between their e-com and old ERP, and rolling out all the pickup/ship options we've discussed. But not everything was high-tech. We also did low-tech wins like posting consistent BOPIS signage at store entrances (so customers knew exactly where to go), and installing those colored light towers to alert staff of online orders. We even issued different colored vests for the "Omnichannel specialists" in stores so it was clear who was handling pickups and curbside. The point is, **omnichannel retail is equal parts technology, process, and people management**. The playbook is: start with the **customer journey** (how do they want to shop, receive, return?), implement the core capabilities one by one (pickup, ship-from-store, etc., per our five models), and continually fine-tune operations on the ground. Use your store network as a competitive weapon – they are closer to customers than your warehouse and can offer human service that pure online players can't. If you empower them with the right tools and data, your stores become a huge advantage in the omnichannel game.

ARCHETYPE 3: PURE-PLAY DIGITAL

This is a business that started online and has no (or minimal) physical retail presence of its own. Think direct-to-consumer brands, e-commerce pure-plays, marketplace sellers, etc. At first glance, you might say "Omnichannel doesn't apply – we're online-only." But even pure digital players are finding ways to create *physical touchpoints* or multi-channel experiences. Why? Because customers still live in an offline world too, and having an omnichannel strategy can set you apart (and even reduce digital customer acquisition costs by boosting word-of-mouth and brand presence). **Quick-Start Moves:** If you're pure online, consider **partnering or piggybacking on others' channels**. For example, many D2C brands have started to appear as "shop-in-shop" or wholesale partners in brick-and-mortar stores (think Casper mattresses in Target, or Dyson demo zones in Best Buy). You don't need your own fleet of stores; a strategic partnership can give you physical exposure. As an MVP, identify one region or city where you have a concentration of customers and do a **pop-up store or event**. It could be a weekend booth at a local market, a rented space for a month, or even a branded Airstream trailer that tours around (Warby Parker famously did this early on). The key is to **bring your digital brand to life offline** briefly and gather insights. Ensure you have a way to capture data – e.g., have visitors sign up with their email or check-in via your app for a discount – so it connects back to your unified customer data. A pure-play can also leverage **alternative "stores"**: for instance, some online brands allow pickups and returns at UPS stores or use Amazon lockers. Those aren't your stores, but they extend your touchpoints. See if logistics partners offer a pickup drop-off integration (many 3PLs now have options to ship to local pickup points). **Tech for Physical Engagement:** If you do dip toes into physical, you don't necessarily need traditional retail POS systems. When we built a mobile POS for Family Farm & Home's curbside, it doubled as a pop-up store register – it was essentially an iPad running a web app tied into their e-commerce backend. It allowed them to do outdoor tent sales and trade shows with full access to online inventory and easy checkout. As a pure-play, you can similarly use **tablet-based POS linked to your online store** to sell in-person without investing in a lot of hardware or separate software. It's all about *experiments*. One week, try selling on Instagram Live (that's a "channel" too – live commerce). Another, try a local delivery partnership to offer same-day in a metro area. The omnichannel mindset for pure-plays is **"be wherever your customer needs you, and make it seamless."** If you primarily sell on a marketplace (say eBay or Amazon), think about expanding to your own site or vice versa, and ensure branding and service are consistent. **Metrics to watch:** customer acquisition cost vs lifetime value in each channel – often an omnichannel presence can lower CAC because your brand visibility increases. Also track overlaps: how many of your customers touch multiple channels (e.g., follow you on social, buy on your site, and see you at an event). Those multi-touch customers are gold; nurture them. I've found that even one physical meetup (like a pop-up store visit) can significantly increase a customer's loyalty to an online brand – it humanizes the brand. So, pure-plays shouldn't ignore omnichannel; they should redefine it to fit their model. *Omnichannel doesn't always mean "own a store" – it means giving your customers multiple ways to engage with and buy from your brand, in a cohesive way.*

Having looked at these archetypes, you might see your business as one of them or a mix. The main idea is to apply the principles in a way that fits your context. Now, let's talk about the technology underpinning all this – the **tech stack ladder** that enables omnichannel at different stages of maturity.

To execute omnichannel strategies, you need the right tech stack – but it's not one-size-fits-all. I visualize it as a ladder of capabilities. You can start on the lower rungs and climb up as your needs grow. Here's a simplified ladder from basic to advanced:

Level 1: Unified Inventory and POS (Foundation). At a minimum, you need a way to keep track of inventory and sales across channels. For a small retailer, this might be as simple as using a single system (like Shopify or Square) to run both your online store and point-of-sale. That gives you one inventory pool and prevents the nightmare of selling the last item twice. If you're larger and already have separate e-com and store systems, invest in integrating them or use an inventory sync tool so that online and offline draw from the same stock data. The tech focus at this level is getting a **"single source of inventory truth"** (back to Pillar 1) and a basic **order notification system for stores** (even email alerts) to handle pickups. Many mid-sized retailers start here by building a custom bridge between their e-commerce platform and their store ERP/POS. That's exactly what we did with the farm & home retailer – bridging Adobe Commerce to an Epicor ERP via flat-file so the website knew each store's stock. It wasn't glamorous, but it worked. The KPI for success at this stage is near-real-time inventory sync and no double-sell incidents.

Level 2: Order Management System (OMS) & Store Fulfillment Tech. As volume and complexity grow, you'll likely need a dedicated **Order Management System**. An OMS is the brain that decides, *"This online order will be shipped from X or ready for pickup at Y,"* and manages the statuses. At small scale, you can do without (your e-com platform or a person can do it). But once you have dozens of stores or multiple options, an OMS (whether a module or a standalone product) becomes invaluable for routing and tracking. We got to this stage with Family Farm & Home – when their SaaS order management couldn't handle curbside, we **built a lightweight headless OMS** ourselves to fill the gap. It sat between the website and stores, orchestrating orders and providing a simple admin UI for store staff to mark orders as picked, packed, or ready. In tandem, equip your stores with *fulfillment tech*: handheld devices or tablets for picking orders, label printers for ship-from-store, scanners to make sure the right items are picked. You don't need to boil the ocean here – start with one tablet and one printer per store as a pilot and see how many orders you can process efficiently. The OMS should capture metrics like pick/pack times per store, so you can optimize. A big part of Level 2 is **training**: the tech won't help if staff don't use it right. One client of mine gave store associates a mobile app but found they still printed the orders out and manually checked items – because the app was clunky. We realized we needed to simplify the UX (we added a "Pick All" one-tap button to confirm all items, instead of 99 separate taps). That change was huge – it came from listening to a frontline cashier who said the old way was killing her efficiency. Moral: build or customize your tech to **mirror store workflows**. At Level 2, your KPIs are things like fulfillment SLA adherence (e.g. 95% of pickup orders ready within 2 hours), store order throughput, and order accuracy.

Level 3: Unified Customer Data Platform & Personalization. Once your transactional systems are humming, the next rung is leveraging all that data to improve the customer experience and marketing. This is where a **Customer Data Platform (CDP)** or some form of unified CRM becomes important. The goal is to consolidate data from e-commerce, stores, email, mobile app, etc., into one profile per customer. Many companies attempt this via loyalty programs – encouraging customers to identify themselves in every channel. Technologically, you might integrate your e-com database, POS customer records, and email service provider into a CDP that can ingest and match users (often by email or phone number). With unified data, you can do neat things: for example, send a follow-up email referencing the exact store someone bought from ("How do you like your purchase from our Dallas store? Leave a review!"). Or train your customer service reps to see a full history: "I see you bought this online and that in-store, how can I help?" On the analytics side, unified data enables **omnichannel attribution** – understanding the true paths to purchase (maybe Google Ads drive online orders while Instagram drive in-store visits). **Tech note:** This level often involves either extending your existing systems (some e-com platforms have modules to capture in-store data) or implementing a CDP like Segment, Tealium, or similar to pipe data from various sources. The caution here is privacy and data cleanliness – ensure you have customer consent to use their data across channels and keep that data secure. **KPIs at this level:** customer retention, repeat purchase rate, and average spend per customer are big ones. Ideally, you want to see omnichannel customers spending more (you likely will, as studies suggest). Track engagement metrics like cross-channel usage (e.g., % of customers who engage both online and in-store). If those numbers go up, it means your unified experience is resonating. Personalization is a key outcome of unified data – your marketing campaigns can be smarter (targeting store-preferring customers with store events, etc.), and even your website can change content based on offline behavior. This is advanced but very powerful; many retailers are still working up to this.

Level 4: Advanced Optimization (AI & Automation). At the top of the ladder, you have a true **unified commerce platform** operating, and you start layering in advanced capabilities: predictive analytics, AI-driven decisioning, and automation to squeeze out every efficiency. For instance, using AI to **optimize inventory distribution** – predicting which store should get more of product X because that region's online demand is spiking. Or dynamically routing orders: maybe an AI engine decides on the fly whether it's more cost-effective to ship from the DC or from a store or even directly from vendor, given the current shipping rates and stock levels. You could also implement **AR/VR experiences** linking online and offline (like AR fitting rooms that drive in-store sales – Rebecca Minkoff is achieving 65% higher conversion after AR tryouts). Another frontier is **automating customer service across channels** – e.g., a customer starts a return online, decides to drop it at store, and your system already knows the details when the return hits the POS. AI can assist associates by providing recommendations ("This customer might be interested in these items, based on their profile"). On the fulfillment side, robotics or at least optimized batching might come into play for in-store picking (some retailers are testing things like automated pickup lockers, or picking robots in backrooms). **Tech note:** at this level, your stack might include an advanced

OMS or distributed order management with AI optimization (some providers or in-house solutions), a CDP with machine learning for segmentation, and IoT devices in stores (smart shelves, sensors). It's bleeding edge stuff. But don't let the shiny toys distract you – none of this works without the solid foundation of levels 1–3. I often have to remind clients who jump to "can we use AI to do X?" – first make sure your data is correct and your operations are smooth. AI on bad data or broken processes just amplifies the mess. That said, once you're ready, these tools can yield new efficiencies and experiences that truly differentiate your brand. **KPIs:** Here you look at things like forecast accuracy, reduction in stock-outs due to predictive moves, increase in customer lifetime value from personalization, etc. Also cost metrics: have you reduced the cost per order by using stores vs DC optimally? Amazon is the king of this (automating and optimizing everything behind the scenes), but mid-market retailers can absolutely borrow these tactics at an appropriate scale.

Wherever you are on this ladder, remember that the goal is **not to implement tech for tech's sake**, but to enable the business outcomes – better customer experience, higher sales, healthier margins. One of my favorite moments was when a client's CIO asked if they should buy XYZ omnichannel suite, and we realized we could meet 80% of their needs by connecting two systems they already had and doing a bit of custom dev – saving them a million dollars and a year of implementation. Always ask: *can we do this in a simpler way?* Often the answer is yes, especially now when APIs and cloud services make integration faster.

THE ESSENTIAL OMNICHANNEL METRICS DASHBOARD

To know if all this is working, you need a clear view of the numbers. I recommend creating a simple **omnichannel dashboard** accessible to all key teams, showing a mix of customer experience metrics and operational metrics. Here are the all-star metrics I always track, and why they matter:

Inventory Accuracy (%): This measures how correct your inventory counts are (by location, ideally). It's usually calculated by periodic audits or cycle counts: (counted stock vs system stock). As we discussed, without high accuracy, omnichannel falls apart. Aim for *at* *least* 95% accuracy in stores; many retailers start far lower (the average is ~65%–70% accuracy, which is alarming). Improving this number directly reduces out-of-stock issues and canceled orders. We tracked this for Rural King and drove it up through frequent syncs and auditing – you should too. **Source:** audit results, RFID systems, etc.

Online Order Cancellation Rate: What percent of online orders are cancelled due to inability to fulfill (usually because the item wasn't actually available or missed SLA)? This is a red-flag metric. Nordstrom Rack's cancellation rate on BOPIS was reportedly very high, which is why they had to scale it back. You want this as low as possible, ideally under 2-3%. If it's higher, find out why (inventory oversells? store pick failures?) and tackle those issues. This metric holds teams accountable for Pillar 1 – inventory truth. **Source:** OMS or e-com platform reports.

BOPIS & Curbside Order Count (and % of Total Sales): Track how many orders are coming through these omnichannel services and what share of overall sales they represent. This shows adoption. If BOPIS is 10% of your e-commerce orders and growing, that's a good sign – customers value it. Target and others have seen huge shares (Target said over 50% of digital sales were fulfilled through same-day pickup/drive up by 2022). It's also a planning figure – higher uptake may mean you need more staff or space dedicated to these services. **Source:** e-com analytics, OMS.

Average Pickup Time / Wait Time: For BOPIS, measure the time from order placement to ready-for-pickup notification (how fast you prepare orders). For curbside, measure how long customers wait in the parking lot. These are customer experience gold. If you promise pickup in 2 hours but average is 4 hours, fix it. We strived for under 1 hour for BOPIS in many projects (for in-stock items), and under 5 minutes wait for curbside arrival. Speed is a competitive advantage – remember, *customers choose omnichannel for convenience.* **Source:** timestamps in your order system and perhaps curbside check-in logs.

Store Fulfillment Rate & Speed: If doing ship-from-store, monitor what portion of online orders are fulfilled by stores, and how quickly. Also, crucially, track any *hand-off delays* (order assigned to store but not shipped out in promised time). This helps gauge if stores are effectively acting as fulfillment nodes. **Source:** OMS.

Attachment Rate (Pickup upsell rate): We touched on this – what percentage of customers who come to pick up an order end up buying something else in that visit? You can measure this by checking if a customer ID (or email/phone) who had a pickup on a given day also made a purchase through the register that day. An increase in this metric is a strong indicator that omnichannel is driving incremental in-store revenue. Some companies have seen 20%+ of pickup customers make additional purchases – that's significant. **Source:** POS transaction logs matched with pickup records.

Customer Satisfaction (Omnichannel NPS or CSAT): It's wise to specifically measure customer satisfaction for your omnichannel experiences. Send a quick survey after a pickup or curbside: "How was your experience?" Track that Net Promoter Score or CSAT separately from general store or online scores. This helps pinpoint if issues exist. For example, if your overall CSAT is 90% but BOPIS CSAT is 70%, you know the new service is underperforming expectations and needs attention. **Source:** post-pickup surveys, feedback forms.

Omnichannel Customer Lifetime Value: More strategic, but extremely telling. Calculate the LTV (or 1-year spend, etc.) of customers who engage in multiple channels vs those who don't. HBR's study showed omnichannel shoppers are more valuable. Your own data can quantify this for your execs (e.g., "Customers using store pickup have 1.5x the annual spend of online-only customers"). This reinforces the ROI of omnichannel initiatives. **Source:**combined purchase data per customer.

Margin Impact Metrics: Since we're calling omnichannel a margin battleground, include a metric or two to monitor cost vs benefit. For example, *Fulfillment Cost per Order* (and break it down by channel – store-fulfilled vs DC-fulfilled vs curbside). Also *Margin per Order* by fulfillment type. If you can show that after some optimization, your store-fulfilled orders have only a 1% lower margin than DC orders (or even higher margin due to saved shipping), that's a win – and vice versa, if you see a gap, that flags an area to improve. Recall that with good systems, stores *can* fulfill profitably (even +3% margin lift). Monitor it.

Channel Migration & Retention: Keep an eye on how customers move between channels. Are your stores driving people to the app? Is your app driving people to stores? Metrics like the number of store customers who later buy online (after maybe being introduced in store), or vice versa. It's less of a KPI and more of an insight generator, but it tells you if your omnichannel approach is holistic. Ideally, you want a high crossover – that means your channels support each other rather than operate in silos.

Seeing these numbers regularly keeps everyone focused on what matters: **serving customers well and doing it efficiently across channels**. If a metric dips, it sparks quick investigation (e.g., cancelations spiked this week – did an inventory sync fail?). If a metric soars (attachment rate up), you can double down (maybe due to a new promo at pickup).

We've covered a lot of ground – from strategy and anecdotes to tech and metrics. The last thing I want to leave you with is an action plan. Talking and planning are great, but *doing* is what creates momentum. Here's a quick **7-day sprint plan** to notch an omnichannel win *this week*.

7-Day Quick Win Sprint

Day 1: Strategy Sprint Kickoff – Gather a small tiger team (e-com lead, store ops rep, IT rep, customer service). Pick **ONE** omnichannel pilot to execute in the next 7 days. It could be as simple as enabling in-store pickup at one location for one product category, or setting up a temporary curbside station for the weekend. Keep the scope narrow. Today, clarify the goal (e.g., "Allow customers to pick up online orders at our Downtown store") and outline the tasks for the week. Secure any necessary manager buy-in by promising it's just a test (and highlighting how it could drive extra sales).

Day 2: Inventory & Systems Check – Make sure you have the data you need to support the pilot. If it's BOPIS for certain items, ensure you know the store stock of those items *right now*. If needed, do a quick manual count of 20 top SKUs – verify system accuracy. Update any inventory files or system settings to "turn on" those SKUs for pickup at the pilot store. Also, configure your e-commerce front end: create a simple banner or note on those product pages – "Pick up in store available!" – so customers know. If your platform allows store selection and shows availability, great. If not, a basic workaround is to list those items as available at that store in a text list or FAQ. Not pretty, but workable for a week test.

Day 3: Process & Training – Head to the pilot store (or video call with the staff if remote). Walk through the **entire pickup process flow** with the store manager and a couple of associates. For example: *"Customer orders online, you'll get an email with order details, here's how you acknowledge it. Here's where you'll put the items (let's clear a shelf near the counter). We'll send the customer an email to come pick up once you confirm the item is set aside. When the customer arrives, greet them, verify ID or order number, hand over the goods with a smile and maybe an upsell suggestion."* Keep it simple, but make sure store staff know their roles. This is *crucial* – many pilots fail due to frontline confusion. Print a one-pager cheat sheet for them. Also decide on a *communication channel*: maybe a Slack channel or text group for the team to quickly address any issues that come up during the pilot.

Day 4: Go Live (Soft Launch) – Quietly flip the switch. Make the option live on your website for that store (or have your customer service offer it to any order that would make sense). On day 4, perhaps only inform a few friendly customers (or internal folks) to test it out. You might even have an employee place a mock order online to go through the motions. The point is to ensure the whole pipeline works: order comes in, store is notified, item is picked and held, pickup notification goes out, etc. Monitor closely and iron out any bugs. Did the email notification go out correctly? Could the store staff mark the order as ready easily (even if that's just sending a manual email for now)? Fix anything broken while volume is low.

Day 5: Public Announcement – If all looks good with the dry run, announce the pilot more broadly. Put out a social media post or email targeted to customers near that pilot store: "We're testing Buy Online, Pick Up In-Store at our Downtown location! Be the first to try and skip the shipping wait." Often, customers are excited to be part of something new. Make sure customer service knows about it, so they don't look clueless if someone asks. Orders may start to trickle in. Today you'll start getting real data – watch it. If an order comes and the store doesn't confirm in, say, 1 hour, give them a nudge – maybe they got busy, and that's okay, but this is learning what realistic timing is. By end of Day 5, you hopefully have a few real pickups completed or scheduled.

Day 6: Observe & Gather Feedback – Spend time at the store if possible. Watch the process in action with actual customers. Talk to the store staff – what's working, what's awkward? Maybe the email notifications are too sparse in info, or the customer came to the wrong counter because signage was unclear (slap up a quick printed sign if needed!). Also maybe call or survey the customers who used it: *How was the experience?* Did they find it easy? Nothing beats direct feedback at this stage. Also look at sales data – did any of the pickup customers buy extra items? Even anecdotal evidence ("Customer X grabbed another item while here") is valuable. Use today to document these insights. Fix any minor issues you can immediately (agility is the beauty of a quick pilot).

Day 7: Celebrate & Evaluate – You did it – you stood up an omnichannel service in a week! Take a moment to high-five the team (maybe literally, if you're at the store). Small wins build momentum, and showing that you can move fast breaks down internal resistance. Now evaluate: How many orders came through? If it's just a handful, that's fine – you learned a ton. What went well and what needs improvement if you were to expand this? Perhaps the tech held up but the store is understaffed at peak hours for this – note it. Or customers loved the convenience but one complained it took too long – note that too. With this week's data, you can make a go/no-go decision on scaling the pilot further. In most cases, a modest success is enough to justify expanding to a few more stores or products. Importantly, share the success with your broader team and executives. Show them pictures of the pickup shelf, share a customer comment ("This is awesome, thank you for offering store pickup!"), and highlight any extra sales captured. This turns omnichannel from an abstract idea into a tangible reality for your business.

Seven days, one win. It may be small, but it's progress. As I often coach: *momentum matters*. Once you prove you can launch something quickly, it builds confidence and appetite for more. Maybe next week you'll try curbside, or add another store to BOPIS, or sync an inventory feed to Google so local shoppers see what's in-stock (that alone can drive a big lift – one retailer I know saw a 25% jump in store pickups after feeding store inventory to Google). The key is to keep iterating and *learning by doing*.

Congratulations on making it through this omnichannel deep dive. You've seen what works, what fails, and how to get started without boiling the ocean. The competitive advantage in 2025 goes to those who can execute quickly and delight customers seamlessly across channels. Be that retailer.

Next up: **Chapter 3.1 – Tech Debt: The Silent Killer of Margin.** Now that the front-end channels are firing, we're going under the hood to expose the cruft dragging your profit down: outdated code, half-baked integrations, and quick fixes that became permanent. We'll rip into real-world examples, show you how to spot the invisible interest ticking away every sprint, and lay out the fastest ways to pay it down without stopping the bus. Gear up—this is where we start **building an engine that scales**.

3.1: TECH DEBT - THE SILENT KILLER OF MARGIN

Brrr... It's 6:00 AM on Black Friday and my phone jolts me awake. On the line is a panicked Chief Marketing Officer. Her company's big Black Friday promotion – the one they've been planning for months – isn't working. Customers are flooding social media with complaints that the promo codes don't apply at checkout. Revenue is free-falling by the minute. I log in to help triage and quickly uncover the culprit: an *old, unmaintained piece of code* running their promotion engine. It was written as a "temporary" work-around years ago by a previous developer and never revisited. In that early-morning firefight, one thought hits me hard: **technical debt is like dark matter – you can't see it, but it warps every revenue curve on your dashboard**. We had never seen this issue in tests, but its invisible gravity was bending their results out of shape at the worst possible time. This is the silent killer of margin that hides in the dark corners of your ecommerce stack.

THE HIDDEN COST OF SHORTCUTS: WHAT IS TECH DEBT?

Let's start with a simple definition. *Technical debt is the hidden cost of shortcuts and work that needs to be done, but hasn't.* It's the accumulation of all those quick-and-dirty fixes, rushed launches, skipped upgrades, and "we'll refactor it later" decisions we make in order to hit a deadline or save budget. In the moment, taking a shortcut can seem harm-less – even necessary. But over time, those deferred improvements accrue interest in the form of slower performance, fragile code, and lost opportunities. Eventually, the "interest payments" (like extra mainte-nance work or firefights during Black Friday) come due with a vengeance.

It's important to understand that tech debt is **not just "bugs" or broken code.** In fact, you can have zero obvious bugs and still be drowning in tech debt. I've seen ecommerce sites where certain features quietly don't work as intended, but the team has normalized the problem as "that's just how our site works," when in reality it's a red flag for deeper issues. Tech debt isn't just a developer inconve-nience – it's a business problem that **impacts revenue, customer expe-rience, and your ability to compete**. That unmaintained promo engine from the Black Friday incident is a prime example: it wasn't throwing errors in QA, but it silently failed when the stakes were highest.

In my experience, there are three main types of technical debt in ecommerce:

Intentional Debt: This is when you *consciously* take a shortcut. Maybe you launch a feature quickly with the plan to refine it later, or you hard-code a fix because you can't miss the holiday launch. You *know* you're creating debt, but you accept it as a trade-off for speed. For example, a team might skip writing unit tests to meet a Black Friday deadline – that's intentional tech debt.

Accidental Debt: This happens when decisions that seemed fine at the time turn out to be suboptimal in hindsight. Often it's due to lack of experience, knowledge, or changing standards. Maybe a developer wasn't aware of a security best practice, or a feature was built in a way that later conflicts with another. Nobody *meant* to create a problem, but one emerged anyway.

Aging (or Environmental) Debt: This is the slow creep of outdated technology. Even if everything was done right initially, over time your platform, libraries, and infrastructure age. The world changes around your code. A classic example is an e-commerce site still running on an old PHP or Node.js version well past its prime – the code works, but it's becoming legacy by virtue of time. I call this *"time-induced"* tech debt.

A big myth to debunk is *"debt = bad coding."* In reality, you can have very well-written code that still creates tech debt if it's solving the wrong problem or duplicating what the platform already offers. Imagine you hired an agency to build a custom promotions module because you didn't know the native one could meet your needs if you used it a different way. If that custom module isn't maintained, you might later find out it *silently* conflicts with native features (like our Black Friday horror story). That's debt – even if the code quality itself is decent – because it's a **shortcut around doing it the "right" way (using the platform's native feature)** and it carries a hidden cost. As a rule: if it's not built to last, it's tech debt.

A $1.5 Trillion Black Hole: Tech Debt by the Numbers

Still think tech debt is an "IT problem" you can safely ignore? Think again. By the latest estimates, technical debt costs businesses in the United States about **$1.5 *trillion* a year**. Yes, that's trillion with a "T" – an almost incomprehensible drag on margins across the economy. In surveys, **91% of CIOs say tech debt is their biggest technology challenge**, and **80% of organizations have had to delay or cancel a critical project *because* of technical debt**. When we talk about tech debt being a silent margin killer, this is the evidence – it's siphoning off time and money at a massive scale.

Let's zoom in further. One research study quantified the cost of tech debt at **approximately $306,000 per *million lines of code*, per year**. In other words, for every million lines in your codebase, you might be sinking a quarter-million dollars annually into dealing with the issues caused by debt. That figure equates to roughly **5,500 developer hours wasted** on rework, patching, and maintenance that wouldn't be necessary if the code was cleaner. Think about that: 5,500 hours is the labor of nearly three full-time developers for a year, *per million lines of code*, just fighting fires and shoring up shaky infrastructure.

Now, you might be saying, "Sure, big national number – but what about my business?" Let's make it tangible. Suppose you run a $50 million-a-year online store where most of your sales occur between 8 AM and 8 PM. That's an average of about **$190 of revenue per minute**.

Even just a ten minute glitch during peak hours could cost you almost **$2000** – and that's just the direct hit, not counting the customers you lose.

To put it plainly, technical debt isn't just an abstract code quality issue. It has a **real, monetary impact on your business**. It acts like a black hole on your P&L – an unseen force pulling your margins down. You'll see it appear in sneaky ways: a sudden spike in bounce rates due to slowing pages (lost revenue), an expensive emergency upgrade project you had to fund to patch an unsupported system (unplanned expense), or the opportunity cost of features your team *couldn't* deliver because they were busy firefighting.

Let me give another example: we worked with a mid-market retailer that used to schedule big site updates during "maintenance windows," often taking the site offline for hours overnight. They eventually realized these maintenance-mode periods were costing them about **$5K in lost sales per minute, even during overnight deployments**. By tackling their deployment process debt (more on that soon), we helped them shrink those blackout windows from hours to just seconds – essentially eliminating the revenue loss – and *accelerated their new feature launches by 300%* in the process. That's the very real ROI of paying down tech debt.

THE DEPLOYMENT PARADOX: SLOW RELEASES = MORE DEBT

One of the most counter-intuitive aspects of technical debt is what I call **the deployment paradox**. It goes like this: teams accumulate a lot of tech debt, which makes them *fearful* of deployments ("If we push an update, everything might break!"). So they slow down their release cadence to play it safe. Instead of deploying daily or weekly, they start batching changes into big releases once a month or once a quarter. That approach is meant to *reduce* risk – but ironically, it often ends up **creating more risk and more debt**.

Here's why: when you delay deployments, you're not avoiding change; you're *stockpiling* it. Changes still happen – requirements evolve, customers demand new features – but now they pile up in a long queue. By the time you finally deploy, it's a monstrous release with a ton of moving parts. Testing becomes harder, the deployment process grows more complex, and if something goes wrong, it's *really* hard to pinpoint the culprit among dozens of updates. So you've traded lots of small, manageable risks for one colossal, scary release. And because it's so scary, what do teams do? They delay it *further*, trying to test and perfect it – which only makes the next release even bigger. It's a vicious cycle, and your tech debt *keeps growing in the background* as you hesitate to touch anything.

Meanwhile, consider the companies that have broken out of this trap. **Netflix deploys code 4,000+ *times per day*.** That's not a typo – thousands of production deployments a day. Amazon famously deploys every 11 seconds on average. While you're sitting on your quarterly "big release," companies like Netflix, Amazon, and Walmart are shipping dozens of features and fixes *per hour*. They're not doing this because they enjoy risk – they're doing it because **small, frequent, automated deployments are actually *safer***. Each change is tiny, isolated, and quickly rolled back if there's an issue. There's less chance for unintended side effects to creep in. It's like steering a speedboat versus an oil tanker; the speedboat can make quick adjustments to avoid danger, whereas the tanker needs miles of warning to change course.

I often tell folks: *the businesses that grow the fastest tend to iterate faster than their competition*. There's a direct line between deployment frequency and business agility. If you can only deploy new features once a quarter, you have at most four shots a year to improve your customer experience or react to market changes. Your competitor deploying weekly has 52 shots. Who do you think learns faster and adapts quicker?

Now, moving from a slow release cycle to an agile one is easier said than done – especially when you're carrying a backpack full of tech debt. But there are two game-changing practices that can make this transition far more achievable: **feature flags** and **continuous deployment**.

Feature flags are one of my favorite tools for breaking the deployment paradox. They're basically exactly what they sound like: a way to put a "flag" in your code that can turn features on or off dynamically. In practice, you wrap new or risky code in a conditional that checks, say, a config setting or database value – *if the flag is ON, execute the new* *code; if OFF, skip it.* This means you can **deploy code to production in a turned-off state** and enable it when you're ready. The beauty is that you separate *deploying* code from *releasing* a feature. The code can be sitting harmlessly in production, doing nothing until you flip the flag. When you're ready, you might turn it on for, say, 5% of users as a test. If something goes wrong, you flick the flag off – *no rollback deploy, no downtime*. With feature flags in place, the user never experiences any downtime during a rollout.

Continuous deployment (CD) goes hand-in-hand with this. CD means automating your release pipeline so that code can flow from development to production rapidly and reliably. Automated tests, integration checks, and monitoring are the gatekeepers instead of manual human approvals that take days or weeks. Embracing CD doesn't mean you throw caution to the wind – it means you build *quality checks into the pipeline* so that you can deploy fast *with confidence*. You might deploy small changes dozens of times a day, behind feature flags, with each deployment being no big deal.

Consider this contrast: I've worked with new clients who were terrified of deploying – their process was essentially "everyone hold your breath and pray" on the big day. They'd do it maybe once a month, overnight Sunday, with ops and developers all on high alert (and high stress). Compare that to a company like Netflix where deployments are so routine they're boring; engineers deploy whenever their code is ready, and they've got tooling to automatically halt or roll back if any anomaly is detected. The result? **Less debt accumulates**, because issues are addressed when they're small. And teams aren't afraid to make improvements, because their deployment process can handle change.

The takeaway here: *If fear of your fragile system is slowing down your releases, that's a huge red flag.* Your caution is understandable, but it might be costing you more in the long run. By adopting techniques like feature flags and investing in continuous deployment automation, you can actually decrease risk and start paying down the deployment-related debt. In fact, one mid-market retailer client of ours shifted from infrequent, "big bang" releases to continuous deployment with feature flags – and they **reduced their maintenance-mode downtime**

windows from hours to seconds, while tripling their deployment frequency. They went from being stuck in port like a cargo ship to moving with the agility of a speedboat. And their revenue curve thanked them for it.

Scoring Your Tech Debt: The Creatuity Framework

At this point you might be thinking, "Alright, I know we have tech debt – but where do we *start*? How do we get a handle on this monster?" When *everything* feels urgent and important, it can be paralyzing. I've seen teams with a backlog of hundreds of "debt tasks" just freeze up, unsure what to tackle first. When everything is urgent, when everything is important, then nothing gets done. To cut through that overwhelm, my team at Creatuity developed a straightforward **Tech Debt Scoring Framework**. It's a practical tool to *identify*, *assess*, and *prioritize* your tech debt so you know exactly where to focus first for maximum impact.

Here's how it works in a nutshell:

Identify – First, you need a complete picture of your tech debt. We recommend a two-pronged approach: quantitative and qualitative.

Quantitative identification means using tools to scan your code and systems for red flags. Static code analysis, linter reports, for Magento projects there's even tools like Magento's Test Automation or the community Magento **Test Essentials** packages – these can automate the detection of things like complex spaghetti code, duplicate code, security vulnerabilities, outdated dependencies, etc. They give you a *data-driven* list of potential debt items. For example, a scan might reveal that you have 5000 lines of custom code overriding core checkout functionality – that's a hint to investigate why so much override was needed.

Qualitative identification is equally important. This is the human side: talk to your team, do interviews, and use your intuition from living with the system. In development we talk about *code smell* – places in the code that "smell" wrong to an experienced developer. I believe in ecommerce operations we often encounter **"platform smell."** It's that sense you get that something is off in the platform as a whole. For instance, are there **native features that aren't being used but really should be** (like a built-in promotion or tax calculator that someone replaced with custom code)? Are there parts of the site that frequently crash or that everyone avoids touching because they're so brittle? Document all those. I often find that some of the **most critical tech debt is hiding in plain sight** – things your team has worked around for years ("Oh, we never use the CMS's built-in bundling feature because it doesn't work with our setup"). Those "platform smells" can point to significant debt. In one case, a client's team mentioned they *never* updated a certain extension because it would break the site – huge red flag! It turned out that extension was heavily modified in ways no one fully understood anymore. That's tech debt identified qualitatively.

Assess (Score) – Next, for each debt item you've identified, we **score it across three factors: Urgency, Business Impact, and Complexity**. Each factor gets a rating from 1 (low) to 5 (high). This scoring is the heart of the framework:

Urgency: How pressing is this issue *right now*? Is it actively breaking things or about to cause a crisis, or is it mostly an inconvenience? For example, a bug crashing the checkout daily is a 5 for urgency (it's happening in real time), whereas an outdated library that still works fine might be a 2 or 3 (important to fix eventually, but not burning down the house today).

Business Impact: To what extent does this debt affect revenue, customer experience, or efficiency? Does it hurt conversion rates, slow down the site, cause SEO penalties, chew up team hours, etc.? A debt item preventing promotions from working during sales has a *huge* business impact (score it high). Something like an admin panel glitch that only affects an internal report might be low impact.

Complexity (or Effort): How hard will this be to fix? Is it a quick config change or does it require a major rebuild/refactor? The more complex, the higher the score. For instance, replacing an entire legacy integration with your ERP might be a 5 in complexity – a significant project – whereas removing an unused snippet of code is a 1.

Once you score each item on these three dimensions, you can add them up or plot them. In our downloadable worksheet, we provide a formula that combines the factors (with a bit of weighting) and then we map the issues on a chart. The simple version, though, is you'll start to see what *pops out* as high urgency, high impact, *and* low complexity – those are your low-hanging fruit. Conversely, you'll see which items are low impact, not urgent, and extremely hard to fix – those are likely candidates to consciously defer.

Prioritize – This is where the magic happens: using the scores to prioritize and create an action plan. I like to visualize it as a **four-quadrant grid** (impact vs urgency on the axes, with bubble size or color for complexity). This gives you an at-a-glance map of what to do now versus later. In general, follow these principles which our framework makes very clear:

High Urgency, High Impact, Low Complexity: *Do this immediately.* These are your **quick wins** that will give you a big bang for buck. For example, in the case of the broken promo engine, once we scored it (Urgency 5, Impact 5, Complexity 1) it was obvious this had to be the first thing to fix. One sprint later, promotions were working even under heavy traffic again and margin was back up. You don't wait on a high-high-low item – tackle it now and reap the rewards.

High Urgency, High Impact, High Complexity: This is tough – it's mission-critical but also hard to fix (for instance, a core system that needs a re-platform). These often become major initiatives that you'll break into phases. If something scores high-high-high, you may need to **escalate it to leadership** as a strategic project (because it will require significant time or budget), or find creative ways to reduce complexity (like bringing in an expert or using a modern tool to assist). But don't let the high complexity scare you into doing nothing – if it's truly high impact and urgent, you *must* address it; just plan for how.

Low Urgency, High Impact: These are interesting – not on fire now, but if solved could unlock a lot of value. Many optimizations or refactoring tasks fall here. I suggest scheduling these into your roadmap after the quick wins. Keep an eye that they don't become urgent by festering.

High Urgency, Low Impact: Something like a minor bug that annoyingly throws errors in the logs. It might feel urgent to a developer but if it doesn't really affect the business, you could rank it lower. Fix it in a batch with other small issues when time permits.

Low Urgency, Low Impact: *This is debt you can live with for now.*

Document it, keep it on the backlog so it's not forgotten, but it's likely not worth fixing immediately. For example, an outdated library in a part of the site that's rarely used – note it, but don't let it distract from higher value work. The key is to acknowledge it (so it doesn't become invisible) but consciously decide "not now." As I often advise: not all debt must be paid off *today*; you just need a plan so that you carry it knowingly and don't let it balloon.

By scoring everything, suddenly an overwhelming mess of "issues" turns into a clear picture with priorities. One client remarked that just going through this exercise turned chaos into clarity – the team stopped arguing anecdotally ("I *feel* this is important!" "No, that is!") and started aligning on a data-informed plan. It takes a lot of emotion out of the decision when you can all see on a chart which items will give the best ROI if fixed. And importantly, the framework brings **business context** into technical decisions – it forces you to think about revenue and customer impact, not just code purity.

To illustrate, here's a **sample Tech Debt Scorecard** for a fictional ecommerce site. Each item is scored and the priority becomes evident:

Broken promo code engine– native promotion features disabled by custom code causing discount failures during sales (e.g. Black Friday)

Urgency: 5 – Failing in peak sales now, very urgent.

Impact: 5 – High revenue impact, promos not applied = lost sales.

Complexity: 2 – Low complexity; likely just reintegrate native feature or fix a small module.

Top Priority. Fix immediately (restore native promo functionality).

Random checkout crash – intermittent checkout page errors ("That's just how our site works" syndrome)

Urgency: 5 – Crashes happening daily, hurting orders now.

Impact: 5 – Very high impact; checkout failures = lost orders, lower conversion.

Complexity: 3 – Moderate complexity; requires module rebuild but isolated.

Top Priority. Fix in next sprint (e.g. rebuild checkout module).

Outdated payment gateway module – still using deprecated API, though it functions currently

Urgency: 2 – Not urgent (still working), but a ticking time bomb if the gateway API updates.

Impact: 4 – Payments are critical; failure would be catastrophic, and old module may lack new features (Apple Pay, etc.).

Complexity: 4 – High complexity to update due to API changes and testing needed.

Plan & Schedule. Not an emergency, but schedule an update in coming quarter before it becomes urgent.

Legacy front-end theme – heavy custom UI code from 5 years ago, making redesign costly

Urgency: 1 – Low urgency (site looks okay, no immediate breakage).

Impact: 3 – Moderate impact; site is slower on mobile and UX is dated, hurting engagement slightly.

Complexity: 5 – Very high complexity to overhaul (would require front-end rebuild or re-platform).

Accept for Now. Document this debt and incorporate into long-term replatform/refactor plans. Not worth quick fix; monitor impact (e.g. Core Web Vitals).

In this example, you can see how scoring illuminates priorities. The promo engine and checkout crash are no-brainers – high-high with manageable complexity, so they should be tackled *immediately*. The payment gateway is high impact but not yet on fire; we slate it for a near-future project. The legacy theme is something to keep an eye on but likely address when doing a larger redesign (defer for now). By laying it out this way, you create a to-do list that's aligned with business outcomes, not just technical ideals.

When you use Creatuity's Tech Debt Scoring Framework (we provide links to worksheets and tools for this in the Resources section of the book), you'll be able to systematically knock down the most harmful debt first. As you do, you'll actually start to free up more resources – it's a virtuous cycle. In fact, when companies rigorously follow this identify → score → prioritize → fix process, they often see significant efficiency gains. I've witnessed businesses achieve a **30-40% reduction in maintenance overhead** after a dedicated technical debt pay-down initiative. Think about that – almost a third of the time and money that was being wasted on dealing with issues got re-purposed to new development and growth. The framework gives you a clear action plan and hope: *tech debt is not an endless whack-a-mole game; you can tame it and even turn it into a strategic advantage.*

FIVE QUICK WINS TO START CUTTING DEBT

Tech debt can feel overwhelming, but not all fixes require a months-long project. Here are five **quick wins** I often recommend to ecommerce teams as a starting point to reclaim some margin *fast*:

Restore Native Features: Revisit areas where you've heavily customized or bypassed the ecommerce platform's native capabilities. Often, I find merchants paying the "tax" of maintaining custom code for things the platform can do out-of-the-box. For example, one retailer's Magento promotions had been entirely custom-coded by a past agency, and they never realized the native promo engine could handle 90% of their use cases. The result? Native promotions stopped working altogether due to those customizations. Our quick win was to remove the hack and restore the native feature set – immediately promotions started applying correctly, and the team could use Magento's built-in promo tools with no ongoing code maintenance. If you have features you've turned off or replaced because "they didn't work for us," evaluate *why*. It could be a case of tech debt (like misconfigured or improperly implemented features) rather than an inherent platform limitation. Unused native functionalities are a big signal of tech debt. Use what's there – it's usually better tested and supported than a quick fix was. You'll save on future upgrade pain and probably unlock new capabilities that have since been added to the platform.

Remove Unused Extensions and Code: Your ecommerce platform might be dragging a lot of dead weight in the form of modules, plugins or custom code that are no longer used. Every extension you install or line of custom code you add is another potential failure point (and often adds a bit of performance overhead). If it's not providing value, **prune it**. I've seen sites with over a hundred extensions installed, but only a fraction are actually active – maybe a leftover loyalty program, an old integration that got replaced, etc. That's bloat that complicates every upgrade and can even introduce conflicts. By uninstalling or removing unused code, you simplify the system. One client removed 20% of their extensions in a week (mostly things their team didn't even realize were still there from years ago) and immediately saw page load improvements and fewer weird bugs. This is a quick win because it's usually straightforward: it's easier to delete something than to build something new! Just be sure to follow proper removal procedures (turn it off, test in staging, then uninstall) to avoid surprises. Lean codebases are healthier codebases.

Enable Automated Testing (and CI): This might sound more like a long-term practice than a "quick win," but setting up even basic automated tests can pay back within days by catching issues early. If you don't already have continuous integration (CI) and a test suite running, start with something modest. For instance, Magento (Adobe Commerce) projects can adopt the Magento Test Framework or PHPUnit tests for critical flows. Even a small suite of smoke tests (add to cart, place order, etc.) that runs on each deployment will give you confidence to deploy more frequently. The quick win here is reduced fear – suddenly you're not manually testing everything or rolling dice with each release. It also prevents *new* debt from sneaking in because your tests will flag code that doesn't meet your standards. Implement continuous integration with automated testing to trap new debt before it reaches production. You can start seeing benefits in the very next release – maybe your test catches a bug that would have caused an outage. That's immediate ROI. Over time, you'll expand test coverage, but even a little bit goes a long way.

Add Feature Flags: We discussed feature flags in depth above; here I'm suggesting as a quick win to actually implement a basic feature flag system in your app. There are lightweight ways to do this – it could be as simple as a config file or database table that your code checks, or using an open-source library. Start by identifying one upcoming feature (or a risky fix) and put it behind a flag. Roll it out to an internal team or 1% of users, monitor, then progressively dial it up. By doing this once, you'll gain the muscle memory and infrastructure to do it for future features. It dramatically lowers the risk of deployments, which means you and your team will feel freer to address tech debt (because you can ship changes in small pieces under flags). For example, one retailer implemented a feature flag for a new checkout refactor – they deployed the new code live but invisible. Then they turned it on just for employees. We gathered data, fixed issues, then enabled it for everyone. Zero downtime, zero drama. This quick win is about setting up the *mechanism* of feature flags. Start with basic feature flags to enable safe on/off toggles for new code.

Upgrade Your Runtime (PHP, Node.js, etc.): Running your platform on an outdated PHP or Node version is a silent killer of performance and security – a form of tech debt that's relatively easy to pay down. If you're on PHP 7, plan a jump. These upgrades typically deliver immediate speed improvements (often 20-30% faster script execution from PHP7 to PHP8, for example) and ensure you continue receiving security updates. It's a quick win in that the *effort* is usually well-contained (a one-time upgrade project that might take days or a couple of weeks of work, depending on complexity), but then yields benefits indefinitely. I've seen merchants procrastinate on PHP upgrades until they were forced (end-of-life). But when they finally did it, they kicked themselves for not doing it earlier – their pages got faster and their servers handled more load with the same hardware, literally overnight. The same goes for other infrastructure: update that database version, clear out old OS packages, modernize your front-end build tooling. These don't directly add new features, but they remove the invisible friction that's slowing everything down. And importantly, being on current, supported software versions reduces future tech debt – you won't be scrambling when something reaches EOL. Quick tip: schedule runtime upgrades immediately after peak season, so you have runway to work out any kinks before the next big rush.

Each of these "quick wins" addresses a common source of tech debt pain. They are relatively low cost or low risk to execute and can show tangible improvements within a sprint or two. Equally important, they build momentum. When your team sees that, for example, removing 10 unused plugins made the site a bit snappier and freed us from paying for support on those plugins – they'll get motivated to tackle the next debt item. Quick wins prove that *you are in control* of your platform's destiny, not the other way around. They create breathing room and goodwill to go after the bigger debt issues next.

REAL-WORLD TURNAROUNDS: TECH DEBT IN ACTION

Sometimes the best way to understand the impact of tech debt – and the value of addressing it – is through real stories. Here are a few brief case studies from my own experience that show tech debt's effect on businesses and the results of fixing it:

Broken Magento Promotions – Fixed in 2 Sprints: A mid-market retailer struggled with an invisible problem: none of their **native promotions or advanced pricing rules were working** - all their promotions were being provided by a custom module that no one still at the retailer knew about. They only discovered it during a major sale when discounts failed to apply, crushing their conversion rate. We traced it to a previous agency's code that had **"hacked" the pricing logic** – essentially bypassing the native promo engine in favor of a custom module that hadn't been maintained. This was pure tech debt; the original developers took a shortcut to implement a complex pricing rule, but in doing so they crippled out-of-the-box features. Using our scoring framework, we rated this as high urgency (sales were at stake), high business impact, and luckily low complexity (the fix was to remove or rewrite the custom code and re-enable the platform's built-in promos). It became the top priority. Within **two sprints**, we had **restored the full native promotion functionality**, and the company immediately recaptured lost revenue in the next sale period. The CMO told me it was like "lifting a dark cloud" – their marketing could run all the promotions they dreamed up, without fear the site would undermine them. The lesson: if something in your platform *should* work but doesn't because of some old customization, that's tech debt robbing you blind. Fixing it can deliver an instant boost to your top line.

Pandemic-Era Rebuild Cut Update Time from 100 to 2 Hours: In 2020, a B2B e-commerce company (a school furniture distributor in Texas) found themselves in a crisis. During the pandemic surge in online demand, their site was buckling under tech debt. Page loads were slow, native features (like bulk pricing) were broken, and making even a simple content update felt like pulling teeth – the site was so fragile it took an entire weekend of developer work (easily **100+ hours** of combined effort) to deploy a minor update safely. Instead of a full replatform (which they initially considered, but would be costly and time-consuming), we executed a targeted **rebuild/refactor** over a few

months. We dubbed it the "side-car rebuild" – we kept the core platform but **rebuilt the front-end experience using a modern stack
(headless React)**, offloaded complex B2B logic to a new microservice,
and refactored or removed modules that were causing instability.
Essentially, we paid down the bulk of their tech debt in a focused
project. The outcomes were dramatic: the site's performance improved
immediately, and their deployment process went from that manual
100-hour nightmare to an automated pipeline that takes **about 2 hours
to go live** with a new update. In fact, by modularizing their architecture, we saw roughly a **40% reduction in ongoing maintenance and
new feature development time**. What used to take a team of devs a
week to troubleshoot and release now happens in a single afternoon.
This case shows that even if you inherited a mess (in their case, a
flawed build from an inexperienced agency with lots of technical debt),
you can turn it around without starting from scratch. During the
pandemic, this faster iteration ability literally saved their business as
they could swiftly roll out new online-centric features for their
customers (like bulk ordering for schools doing remote learning). Tech
debt had been strangling them; removing it unlocked their agility
when they needed it most.

Angry Birds: Blue-Green Deploys with Zero Downtime: Here's a
fun one from back in the early days of my career. The official Angry
Birds online store was experiencing massive global traffic and frequent
product launches. Downtime was not an option – if the site went down
even for a minute, thousands of eager fans would be affected. At that
time (around 2012), they implemented a **blue-green deployment** architecture – an advanced solution to obliterate deployment downtime.
They ranservers in two identical environments: a "blue" and a "green"
environment. When it was time to release new code, they'd deploy to
the *inactive* environment (say, green) while the active environment (in
this example, blue) kept serving customers. Once everything on green
passed tests, they flipped the traffic routing instantly from blue to
green. Users experienced a seamless switch – effectively zero downtime, even under huge traffic spikes. The Angry Birds site could do
updates in the middle of the day with nobody noticing a blip. This
approach requires more infrastructure (you need double the servers

running in parallel), but for them it was worth it. The takeaway for our context: if you truly need zero downtime and have the scale to justify it, blue-green deployments are the ultimate tech-debt killer for release risk. It *eliminated* the debt of risky, drawn-out deploys – there was always a backout plan (just switch back to the old version if needed). Few mid-market merchants need a full blue-green setup, but elements of that strategy (like automated rollbacks, staging environments that mirror prod, etc.) can be incredibly valuable. The Angry Birds case is an extreme example of how far you can go – they achieved **zero downtime deploys globally**, which at the time was almost unheard of in ecommerce. It paid off with uninterrupted revenues and a fearless ability to update their site whenever they wanted.

Each of these stories highlights a different angle of tech debt: a native feature crippled by custom code, an outdated architecture overhauled in the nick of time, and a deployment process debt solved with cutting-edge strategy. What they have in common is the *after* picture – in every case, fixing the tech debt led to **faster innovation and recovered revenue**. Broken promos recaptured lost sales in weeks; the B2B rebuild enabled rapid updates and saved costs; Angry Birds' deployment strategy allowed continuous improvements with no revenue interruption. These are the kinds of wins that make tech debt worth tackling. It's not about technical purity – it's about unlocking growth and protecting your margin.

DIFFERENT BUSINESSES, DIFFERENT DEBT: TAILORED PLAYBOOKS

Tech debt doesn't look the same for every company. It often follows the contours of your business model and tech choices. Over the years, I've noticed patterns in how different ecommerce archetypes accumulate debt. Let's talk about three common ones and how to approach their specific "flavors" of tech debt:

Side-Car B2B (ERP-Driven Commerce and Pricing Hacks): Many B2B ecommerce sites run in a "side-car" model to a larger ERP or backend system. The ERP handles product data, pricing, customer-specific terms, etc., and the website is almost an afterthought tacked on. The tech debt scenario here usually involves **heavy custom integration code and workarounds to make the website play nice with the ERP**. For example, I know a mid-market brand that built a custom ecommerce platform in 2011 just to tightly integrate with their legacy ERP. Fast forward thirteen years, and that decision became a millstone: *"every new product page still requires a developer push"* because nothing is truly content-manageable. Their mobile performance never improved beyond 2011 standards (since the ERP integration dictated the architecture), yet now 60-70% of traffic is mobile – leading to abandoned carts and lost sales. The small team spends hours on basic site updates instead of marketing or optimizations, and leadership grew frustrated as competitors outpaced them. This is classic side-car B2B tech debt: **custom hacks for pricing and data sync that fossilize your website**. The playbook: First, *inventory those ERP integration points*. Identify where you have duplicate logic (pricing rules in two places?) or batch processes (like nightly CSV exports). Then, consider modernizing the integration – perhaps move to a real-time API-based sync or use a middleware (like an iPaaS) to handle data transformation cleanly. Sometimes a quick win is enabling the ecommerce platform's native features *in conjunction* with ERP data – e.g., let the website apply promotions on top of ERP pricing instead of overriding pricing entirely. In some cases, I've seen success in **offloading complex pricing to a specialized service** (there are pricing engines that plug into ecommerce). The key is to reduce the dependency on hacks. You might not replace the ERP (that's a huge project), but you can refactor the touchpoints so that the ecommerce side can be more autonomous and upgradable. Also, watch out for "one-off" customizations for specific buyers – B2B sites often accumulate special-case code ("if customer = X, apply discount Y") which becomes unmanageable. Consolidate those into a rules engine or at least document and streamline them. By doing these, our B2B clients have seen **order process speed-ups and far less manual data cleanup**. One client cut their catalogue update time from days to hours by eliminating a convoluted ERP export-import script in favor of direct API sync. The goal for side-car B2B: *make the e-commerce platform a first-class citizen*, not just a bolt-on. That usually means untangling it from the ERP's clutches bit by bit and leveraging modern integration approaches. It's hard to grow sales online if your site is effectively "held hostage" by an ancient backend (*digital Stockholm syndrome*, as I like to call it).

Omnichannel Retailer (POS and Inventory Synchronization Issues): Retailers with brick-and-mortar stores plus an online presence have a unique tech debt challenge: keeping data (inventory, orders, customer info) in sync across systems. A common scenario is an **outdated point-of-sale (POS) system or warehouse management system that syncs poorly with the e-commerce platform**. This often leads to debt in the form of custom jobs, scripts, and even manual processes to bridge the gap. Maybe you have a cron job that uploads a store inventory file to the website every night, or a hack to split online orders to multiple store systems. These patches work... until they don't, and when they fail, you get *overselling*, angry customers, and a lot of firefighting. A telltale sign of tech debt here: *inconsistent data*. The website says an item is in-stock at a store when it isn't, or online orders aren't reflecting in the store's system in real-time. If you hear "Oh, we have an Excel macro that store managers run to update that," alarm bells should ring. The playbook for omnichannel: **unify and automate your integrations**. There's little excuse not to have real-time (or near real-time) inventory synchronization. Nothing sours a BOPIS experience faster than a customer ordering an 'in-stock' item online, only to find the store actually ran out. Avoid this by keeping your inventory tightly synced across all channels. If your POS can't push updates, consider implementing an Order Management System (OMS) or a middleware layer that all channels talk to. One retailer we assisted had major tech debt in their BOPIS (Buy Online, Pickup In-Store) workflow – their ecomm site and store system didn't communicate well, resulting in cancellation rates of online pickup orders above 20%. We helped them integrate an OMS that sat in between. Once a web order was placed, the OMS would reserve inventory in the store in real-time and update the ecomm stock. It also handled transfers if needed. This removed a web of custom scripts (and the associated debt) and brought their BOPIS cancel rate under 2%. Also, don't overlook **customer data sync** – loyalty accounts, purchase history, etc. If store purchases don't sync to online profiles and vice versa, teams often create manual import processes that are fragile. Modern Customer Data Platforms or just a well-integrated CRM can solve that, eliminating a lot of pain. The bottom line for omnichannel: your goal is a *single version of truth* for inventory and customer info. Tech debt accumulates when you have split-brain systems. Consolidate where possible, integrate tightly otherwise. And absolutely monitor any nightly batch jobs – every one of those is a candidate for future failure (prefer real-time). Reducing this debt yields happier customers (no more "item unavailable" surprises) and smoother ops. It's a big part of delivering the seamless experience omnichannel shoppers expect.

Pure-Play Digital (Fast Deployment as a Lifeline): Pure online players (D2C brands - online-only retailers) live and die by their website. For these companies, **innovation cadence** is everything. Tech debt for them often appears as anything that slows down their product iteration. If you're pure-play, you likely have a decent tech team and maybe pride yourself on innovation – but debt creeps in here too, usually in the form of scaling issues or legacy choices that didn't keep up once the startup hyper-growth phase kicks in. A classic example: a successful online brand built its site quickly to go to market (maybe on a monolithic platform with a lot of extensions). A few years and 5x growth later, that same platform is now a tangle of plugins, patches, and performance issues that make every deployment risky (sound familiar?). For pure-plays, I often find the **deployment pipeline and testing infrastructure** is where debt hides. Early on, they deploy with just a couple developers – easy. Later, with a bigger team, they still might not have proper CI, automated tests, or staging environments, and it becomes the Wild West. The result: slowing down deployments or causing more production issues – both of which hurt the business. The playbook here is all about adopting *engineering best practices* to regain speed. Think of it as moving from scrappy startup mode to scalable tech company mode. **Continuous Integration/Continuous Deployment (CI/CD)** is your lifeline. If you can invest in any debt reduction, invest in your pipeline – tests, one-click deploys, feature flags, and monitoring. Pure digital players should strive to deploy small changes daily if not hourly. Many of the world's top tech companies (think Shopify, Netflix) attribute their success to deployment agility. It lets them experiment, A/B test, and optimize conversion funnels relentlessly. If your site is your only storefront, you can't afford to be slow or cautious in improvements, so *tech debt that slows releases is* *especially lethal.* I worked with a subscription box company that real-ized their competitors were deploying new homepage experiments twice a week while they were on a monthly deploy cycle. The culprit was a backlog of tech debt: no test automation, a brittle checkout inte-gration, and some bad code that made deployments fail if not done carefully. We tackled it systematically – introduced a basic test suite and CI, refactored the worst offending code (their checkout Javascript), and containerized the app for consistency. Within 3 months, they went from deploying 1-2 times a month to deploying on-demand (usually multiple times a day). They credited this with a significant lift in conversion rate because they could iterate on user experience faster and catch issues before they went live. So for pure-play ecomm, the motto is *"launch early, launch often"* – if tech debt is preventing that, it's priority #1 to fix. Your competitive advantage is agility; losing that to tech debt is like an anchor dragging behind a speedboat. The good news is that pure digital companies often have the talent in-house to fix these issues once it's made a priority. It might mean pausing feature development for a sprint or two to fix your foundation – but you'll come out much faster. I often challenge pure-play CTOs: measure your **lead time** (from code commit to code live). If it's not in hours or days (at worst), debt is probably the reason. Tighten that up and you've gained a massive edge.

Of course, many businesses are a mix of these archetypes. You might have a bit of B2B, some retail stores, and a thriving online channel all in one. The key is to recognize the symptoms in each area. Where do your biggest pain points lie? Is sales blaming inventory data? Is IT swamped with ERP sync issues? Are developers afraid to deploy? Use those clues to pinpoint what type of tech debt playbook to apply. By tailoring your debt remediation to your business model, you'll tackle the problems that matter most for your margin. One size does *not* fit all – a pure D2C should obsess over CI/CD, while a B2B might prioritize stable ERP integrations first, for example. Know thyself, then fix thyself.

MEASURING DEBT: METRICS FOR YOUR LEADERSHIP SCORECARD

You can't manage what you don't measure. I'm a big proponent of tracking a few key **tech health metrics** on your leadership scorecard (if you use EOS / Traction or any management system, these should be up there with sales and marketing KPIs). By monitoring these, you'll have visibility into whether your tech debt is growing or shrinking, and how it's affecting the business. Here are the top metrics I recommend:

Downtime (Minutes of Uptime vs Downtime): This one's straightforward – track how many minutes (or hours) per week / month your site is fully or partially down, including maintenance mode for deployments. Every minute of downtime is lost revenue and lost customer trust. If you start tackling tech debt in your deployment process, you should see downtime plummet. For instance, moving from big bang releases to continuous deployment with zero-downtime techniques might take your downtime from, say, 120 minutes a month to under 5 minutes a month. We certainly want to track that progress. It's also a great motivator: when the team sees downtime drop to near-zero, it reinforces the value of the improvements. And if downtime ever spikes, it's an early warning that perhaps new debt is creeping in or something needs attention.

Core Web Vitals (Site Performance/Experience Scores): Core Web Vitals are Google's metrics for page performance and user experience (like loading speed, interactivity, visual stability). They are an excellent proxy for front-end related tech debt. Heavy, unoptimized code, old images, render-blocking scripts – all those typical "debt" culprits surface as poor Core Web Vitals scores. Track what percentage of your pages (or sessions) are passing Google's CWV thresholds (good LCP, FID, CLS). If a large portion of your site fails CWV, it's a sign of performance debt that will affect SEO and conversion. As you optimize (maybe remove that old bloated slider library or upgrade to a lighter theme), you should see CWV improve. I recall in one month I had ecommerce problem-solving sessions with several brands and every single one of them was failing Core Web Vitals. By tracking it, you can quantify progress. For example, raise the percent of pages with "good" CWV from 40% to 80%. Also, measure page load times, server response times, etc., but CWV gives a holistic user-centric view. It's a great high-level quality indicator that execs can understand ("Our site experience is getting faster / smoother").

Release Lead Time: This is a metric from the DevOps world – basically, how long does it take for a code change to go from a developer's keyboard to live in production? Measured in hours or days. Shorter is better. If your lead time is measured in weeks or months, that's a big red flag indicating lots of process friction or technical blockers (debt!). Elite tech companies have lead times of under a day for most changes. You should at least know what your current state is. Maybe it's "We deploy every 2 weeks, so worst-case lead time is 14 days; best-case (urgent hotfix) is 1 day." By improving your tooling and burning down deployment debt, you might get that to say 1 day typical, 1 hour hotfix. I suggest tracking "average days (or hours) from code commit to live deployment" as a number. Watch it trend down as you improve. If it trends up, something's wrong (perhaps accumulating new debt, or processes becoming too cumbersome). Release frequency (how often you deploy) is related and also good to track, but lead time encapsulates the speed end-to-end.

% **of Development Capacity Spent on Maintenance:** This metric is about resource allocation. Out of your total IT/development hours, what percentage goes to just keeping the lights on – bug fixes, routine maintenance, support – versus new feature development or improvements? High tech debt orgs often find they're spending 50%, 60%, sometimes 70% of their time on maintenance. That's a silent killer of innovation. If more than half your dev budget is feeding the beast of existing systems, it's time to aggressively pay down debt. Track this as a rough percentage. You might have to approximate (some teams track it via time logs or Jira ticket tags). As you pay off tech debt, you should see maintenance effort go down and capacity for new initiatives go up. One company we worked with was able to reduce maintenance work to less than 20% of dev capacity after fixing a bunch of issues (down from about 50%). That effectively doubled their feature output potential. When you report this to executives, it resonates: "We gained back X developer days per quarter for innovation because we're not firefighting as much."

Open Tech Debt Items (# and $ Impact): This is a bit more subjective, but I encourage maintaining a **living list of tech debt items**, scored and ranked as we did above. The metric to track could simply be the *count* of known debt items, or maybe a weighted score total. You want to see that count going down over time, or at least not growing. If it's growing, it means you're likely accumulating debt faster than you're paying it off (bad sign). If it's shrinking, you're on the right track. I've seen teams make a "tech debt burndown" chart like one would for sprint tasks – very effective in showing progress. Another approach is to estimate the *impact* of tech debt items in dollars (harder, but possible for some issues). For instance, you might estimate "checkout bug costs $50K a month in lost sales" – then track the total "cost of open debt" and see that number decline as you fix things. Even if estimates are rough, it communicates the magnitude to leadership ("We've reduced the revenue at-risk from tech debt by 30% quarter over quarter"). Tying dollars to debt makes it real to non-technical stakeholders.

A quick note: these metrics should ideally be reviewed in your leadership meetings. They provide visibility. If downtime ticks up or release speed goes down, the leadership knows there's a tech issue harming the business, and you can rally support (funding, resources) to address it. Conversely, if you dramatically improve these numbers, it's a cause for celebration – and usually corresponds with improved financial performance too (faster site -> more conversions, more new features -> competitive edge, etc.).

One more metric I sometimes include is **developer happiness or churn related to tech frustrations** – it's qualitative, but high tech debt environments burn out developers. If you do employee surveys, track those sentiments. It's cheaper to fix tech debt than to lose good developers who are frustrated by archaic systems.

To summarize, pick a handful of these metrics and put them on your **EOS Scorecard** or weekly KPI dashboard. It keeps everyone accountable that tech health is a ongoing priority, not a set-and-forget. If you're reducing downtime, speeding up pages, deploying faster, and freeing up more dev time for features – you *will* see a positive impact on your growth and margins. These metrics simply make that work visible.

THE 7-DAY TECH DEBT CHALLENGE: TAKE ACTION NOW

Let's get practical. Talking about tech debt is useful, but *doing* something about it is what moves the needle. I want to issue a challenge – an action plan you can execute in the next **7 days** (yes, starting *this* week) to chip away at your tech debt and build some momentum. It's essentially a one-week sprint dedicated to a quick win. Here's your roadmap:

Identify one high-impact, low-complexity debt item. Gather your team for a quick huddle or brainstorming session. Using the framework above, find *one thing* in your backlog or observations that scores high on urgency/impact but seems relatively straightforward to fix. It could be an obvious bug that frustrates customers, a small speed improvement, or an overdue minor upgrade – whatever would noticeably improve things if solved. Don't overthink it; you likely already know a few pain points that fit this description. Pick one. This is your target for the week.

Score it and get agreement that "Yes, this is the one." Take a moment to roughly score the item (if you haven't already) – confirm it's not going to open a huge can of worms. If it suddenly looks more complex than thought, scope it down. The goal is to **choose a bite-sized debt fix** that can be done in days. Communicate to all stakeholders why you chose this – e.g. "This issue is causing 5% cart abandonment, we think we can cut that in half by fixing it – worth $$X – and we estimate it's a quick fix." Getting everyone onboard and excited is key.

Fix it within the next 5 days. Dedicate the necessary developer(s) to tackle this item immediately, and **treat it like a mini sprint or hackathon**. Clear other low-priority tasks if you can. The idea is to create focus: for this one week, Tech Debt Item #1 is king. If possible, involve a QA or business stakeholder early to help test and verify the fix's impact. Work in the open – let the team know "we're fixing this nagging issue and it'll be done by Friday." The psychological effect of *actively working on debt* is powerful. Make sure to apply good practices: e.g., if it's code, write a quick automated test to cover the new behavior so the debt doesn't recur.

Deploy behind a feature flag (if applicable). If the change has any risk or user-facing impact, use this opportunity to implement or utilize a **feature flag** for safety. For instance, if you're refactoring the checkout flow to eliminate a random crash, put the new code under a flag. You

THE ECOMMERCE GROWTH PLAYBOOK   99

can deploy it on, say, Wednesday but turned "off," then switch it on gradually. This way the deployment itself won't be a high-stress event. You can test the flag with internal users or a small percentage of traffic first. Even if you've never used feature flags, try it for this – it's a perfect low-stakes trial. This step isn't required (if it's truly a back-end fix with no risk, you can just deploy normally), but I highly recommend it to start building your flag muscle. The confidence boost of seeing the fix live (even if only to internal users at first) by mid-week is huge.

Measure the outcome and celebrate the win. After the fix is live (by Day 5 or 6), **measure whatever metric this debt item was affecting**. If you fixed a crash, is the error gone and are conversions up? If you sped up page loads, did your LCP score improve? If you removed an unused feature, did it reduce support tickets or memory usage? Get *some* data or anecdote that shows the positive impact. Then – very important – **share the results with the wider team and leadership.** For example: "We deployed a fix for the promo code bug and our average order value is already trending up this week" or "Site uptime was 99.99% this week thanks to no surprise crashes – up from 99.5%". Even a simple "customers are no longer complaining about X, and our CS team is relieved" is worth noting. Make a bit of fanfare about it. This isn't bragging; it's creating a positive feedback loop. It shows that tackling tech debt has real benefits, and it builds momentum and support for doing more. If you have a weekly team meeting, put a slide in about it. High-five the devs who did the work. In other words, **celebrate the win.**

That's it – a 7-day tech debt blitz to prove to yourself and your organization that *you can start reclaiming your margin from tech debt right now*. No need to wait for next quarter's budget or a full replatform project that takes a year. You can make a tangible improvement in a week or less. And once you do one, you can do another. Maybe you make it a routine: every sprint, allocate ~10% of time to a quick debt win. Or have "Tech Debt Tuesdays" where the team fixes a small item each week.

The important thing is to **start**. Tech debt doesn't have to hold you back. You just need to grab the steering wheel and drive through it one issue at a time, guided by a solid framework and a willingness to act.

Having completed this chapter's journey – from understanding tech debt's nature as the unseen margin killer, through frameworks and quick wins, to real examples and action plans – you are now equipped not just to talk about tech debt, but to do something about it. You might feel like a weight is lifting; that's hope and clarity replacing the dread that often accompanies tech debt discussions. This is how we turn tech debt from a silent killer into a manageable, defeatable foe.

Next Up: Chapter 3.2 – Monolith vs. Composable Migration. In the next chapter, we'll tackle one of the biggest strategic decisions that often arises from tech debt discussions: do you continue investing in your existing monolithic platform or take the leap to a modern, composable architecture? Many businesses struggling with accumulated platform debt ask this question. We'll explore how to evaluate your options, the pitfalls of the "big replatform" versus the incrementalapproach, and how to ensure that a migration (if you undertake it) doesn't just swap old debt for new debt. It's a natural follow-on to what we discussed here, because sometimes the ultimate *resolution* of tech debt is a fresh start – but you have to do it right. **Tech debt may be a silent killer, but armed with the right playbook, you won't be its victim.** You're now in control, and whether you evolve your current systems or migrate to new ones, you'll do so with eyes wide open and a bias for action. Onward to Chapter 3.2 and a future where tech debt is firmly under your thumb, not choking your growth.

3.2: COMPOSABLE VS MONOLITH: A PRAGMATIC FRAMEWORK

A scrappy DTC brand I worked with (let's call them *LuxeBaby* to avoid embarassing them) had a cult following for their limited-edition baby apparel drops. One Thursday at 10 a.m., they launched a flash sale after a TikTok video of their new collection went viral. Within seconds, thousands of eager moms (and unfortunately, bots) swarmed the site. The result? **Total meltdown**. The monolithic ecommerce platform buckled under the sudden load – pages froze, carts emptied themselves, and the site ultimately crashed. Meanwhile, scalper bots swooped in and snatched up the inventory. Frustrated customers were forced to hunt for the products on third-party marketplaces at 2–5x markups. The brand hastily emailed an apology admitting "technical failures," but the damage was done. They hadn't addressed the root cause (or the bot problem), and many loyal customers felt betrayed. Trust eroded practically overnight.

Why did this happen? The monolithic architecture of their platform couldn't handle the traffic surge. Everything – product catalog, cart, checkout, CMS – ran through one tightly coupled system, leaving no flexibility when peak load hit. There was no quick way to scale just one piece (say, the cart service) or implement quick bot mitigation; the whole monolith was an all-or-nothing bottleneck. That flash sale fiasco was a harsh lesson: cool marketing can ignite a fire your tech can't put out. And as the founder told me later, "We basically became a victim of our own success on TikTok." In a world where a single viral moment can bring a flood of business, **the old one-size-fits-all ecommerce engine just doesn't cut it**.

Monolith vs. Composable vs. MACH: Making Sense of the Buzzwords

Before we dive deeper, let's get our terminology straight – no buzz-words without clear definitions. In my podcast and consulting, I often have to pause and define these, so here's the crash course:

Monolithic Architecture (Monolith): A monolith is the traditional "all-in-one" platform – one big application that does everything in your store. All your code and features reside in a single, tightly integrated codebase. Think of platforms like early Magento or Shopify – your product catalog, frontend website, checkout, order management, etc., all bundled together. Monoliths aren't inherently evil (despite the bad rap lately); they can be simpler to start with. But as you grow, the "everything in one box" approach can become a liability. Changes can be slow because any update means redeploying and retesting the entire application, and one bug or integration failure can bring the whole beast down. In short, **a monolith is one big unified engine – easy to install, harder to customize or adapt**.

Composable Architecture (Composable Commerce): Composable means assembling your tech stack from **modular building blocks**, rather than relying on one giant system. I often call it the LEGO approach – you pick the best component for each function (catalog, search, CMS, checkout, etc.) and snap them together via APIs. These components can be microservices, third-party SaaS, or custom applications – the key is they're independent pieces that *compose* a larger system. Because pieces are decoupled, you can swap one out without dismantling everything (just like replacing a single Lego brick). Composable commerce is **pragmatic modularity**: you only use (and pay for) what you need, avoid vendor lock-in, and can evolve your stack piece by piece. It's about flexibility – if your checkout is killing conversion, you can drop in a better checkout service, or if your search is weak, integrate a best-of-breed search engine – all without replatforming your entire store.

MACH: Microservices, API-first, Cloud-native, Headless. It's essentially the guiding principles behind modern composable architecture. Let's break it down:

Microservices: Design your system as small services each handling one business capability (product service, cart service, etc.) rather than one monolith. Smaller codebases mean fewer side effects and faster deployments.

API-first: Every component exposes and communicates through APIs. This makes it easier to integrate or replace pieces, since everything talks in a standard way. Typically in modern composable systems this will be based on a standard called GraphQL.

Cloud-native: Utilize cloud infrastructure and services to auto-scale and manage your applications. Cloud-native systems can grow on demand (no hurriedly buying servers during a traffic spike) and often come as SaaS or managed platforms.

Headless: The front-end (what the customer interacts with) is separated from the back-end logic. You might have a custom website or app that calls APIs from backend services. **Headless** gives you freedom to design any customer experience without being constrained by the backend's templating or front-end limitations.

In short, *monolith vs composable* is like buying a pre-furnished house vs. building one to your specs out of modular rooms. The former is quicker to move into; the latter ultimately lets you design the house any way you want. And *MACH* is the architectural philosophy that makes the composable approach possible at scale, especially for ambitious brands.

THE MID-MARKET URGENCY: WHY IT MATTERS NOW

You might be thinking, "Sure, composable sounds nice in theory, but do I *really* need it? My platform has worked fine so far." I'll be blunt: for many mid-market ecommerce brands (the ones not quite enterprise, but outgrowing small business tools), this is make-or-break time. Here's why:

SaaS Sprawl is Real: Over the past few years, I've seen mid-sized companies enthusiastically adopt SaaS tools for every little feature –

one for reviews, one for loyalty, one for subscription orders, one for search, etc. It's easy to sign up and plug them in. Fast forward and you've got a *spaghetti integration monster*. Each additional SaaS adds cost and complexity, and often they don't play nicely together without some duct tape. The irony is you end up with a de-facto *composable* Franken-stack, but without a coherent plan – multiple admin panels, overlapping functionality, and data siloed all over. Mid-market firms feel the pain in both budget and productivity. If this sounds familiar, it's time to rationalize and get deliberate about composable architecture (instead of accidental, ad-hoc composability).

ROI Pressure and Rapid Change: Mid-market ecom leaders face intense ROI expectations. You don't have the luxury of multi-year IT projects that may or may not pay off. If you invest in new tech, it better move the needle *this quarter* or by end of year. This creates urgency to improve site performance, conversion, and integration efficiency *now*. A composable approach, when done pragmatically, can target specific pain points (e.g. slow checkout, weak mobile UX, manual inventory updates) and fix them incrementally – proving value at each step. In contrast, sticking with a stagnant monolith might mean you're waiting on a big version upgrade or locked into a vendor's roadmap, unable to innovate at the pace your business demands.

Vendor Lock-In Risks: Many mid-market brands are realizing they've bet too much on a single platform or vendor. Maybe you chose a popular all-in-one SaaS platform that was great to launch on, but now its limitations are showing – be it lack of flexibility, rising costs, or an upcoming end-of-life announcement. I had a client in the farm-and-home retail space that learned this the hard way. They were locked into a proprietary order management system (OMS) that couldn't adapt to new omnichannel needs. Training staff on that system was a nightmare, and integration with their newer eCommerce front-end was clunky. We helped them replace that vendor's OMS with a composable solution built on Magento's backend and GraphQL APIs. The result: they escaped the vendor lock, cut ongoing costs, and even reduced training time because the new solution fit naturally into their existing tools. The takeaway? If you're beholden to one vendor's ecosystem, you're one unhappy contract renewal away from trouble. Composable architecture gives you options – you can swap out parts or negotiate harder when you're *not* 100% dependent on one platform for everything.

In summary, mid-market brands are squeezed between the quick-launch but limiting SaaS platforms and the powerful but heavy enterprise suites. The path forward is owning your roadmap: selectively adopting composable components where it makes sense, to avoid the SaaS sprawl and lock-in traps while meeting those quarterly ROI goals. It's about being nimble without losing your mind (or shirt) on costs.

THE 5-QUESTION ARCHITECTURE LITMUS TEST

Not sure if your current platform is helping or holding you back? I use a simple litmus test – five blunt questions every ecommerce leader should ask about their tech stack. Your answers will illuminate whether you can stick with what you have a bit longer or it's time to evolve towards composable.

Performance: Is your site fast and stable even under stress? Be honest – how did your site handle the last traffic spike or big promotion? If page loads crawl or, worst-case, your site crashes when volume surges, that's a red flag. Modern customers won't wait for a slow site, and they certainly won't forgive a checkout that fails. A monolith often has more performance bottlenecks (everything competes for the same resources), whereas a well-designed composable setup can scale critical pieces independently. If performance is lagging and tuning the monolith feels like squeezing a balloon, it might be time to decouple and distribute the load.

Integration: How easily can you connect new systems or channels to your platform? In a monolithic world, integrations often mean plugins or extensions bolted onto your core. I've seen B2B merchants install heavy ERP connector modules directly into their monolithic instance – only to have an update on either side break the whole thing. If adding or updating an integration is a hair-pulling experience (or you avoid it for fear of breaking the site), that's a sign your architecture is too rigid. A composable approach emphasizes API-driven integrations – think lightweight middleware or iPaaS services for things like ERP sync, so failures don't cascade into front-end downtime. Ask yourself: Can we plug in a new service (like a personalization engine or new payment gateway) in days, or will it take months and risk our core site? If it's the latter, you have an integration problem.

Change Velocity: How quickly can you deploy changes or new features? This one is huge. In the early days, maybe deploying once every few weeks was fine. Now, if your competitor or an upstart can roll out improvements daily while you're stuck in testing hell, you're at a disadvantage. Monolithic systems often suffer here because even a small tweak means redeploying the entire application - and running a full regression test on 200,000+ lines of code. I've had to explain to a client why a "15-minute code change" required 4 hours of testing – in a monolith, any change might break something else, so you test it all. In contrast, a composable setup with independent services or front-end can deploy updates to one component without disturbing the rest. If your dev team is afraid to touch parts of the system or your release cycles can't keep up with business needs, that's a litmus fail. Faster change means faster learning and adapting, which is gold in ecommerce.

Total Cost of Ownership (TCO): Is the cost (licensing, hosting, development, and maintenance) of your platform scaling reasonably with your business? Or are you dumping more and more money just to keep the lights on? Monoliths can become money pits as you grow – high licensing fees for enterprise tiers, bigger hosting for the one big app, and costly developers who specialize in that one platform. Plus, the maintenance overhead of testing and fixing side effects as discussed. I've seen clients surprised that moving to a composable model *lowered* their ongoing costs: for example, by carving out a custom OMS microservice, one client cut their Adobe Commerce maintenance hours significantly, because 90% of changes only touched that small service. Another saved on licensing by replacing an expensive monolithic module with a cheaper SaaS alternative. Calculate your TCO – not just now, but projected 2-3 years out. If it looks like a hockey stick, you need to rethink your architecture or vendor choices. Composable done right can bend that cost curve down by letting you pay for only what adds value.

Team Skillset and Morale: Do you have (or can you readily acquire) the technical skills to support a more complex, composable

THE ECOMMERCE GROWTH PLAYBOOK   107

architecture? And how does your team feel about your current platform? This is a litmus test that's often overlooked. If you have a small dev team that only knows, say, PHP, throwing Kubernetes, React, and a suite of microservices at them might backfire. On the flip side, if your engineers are itching to use modern frameworks, or you have a strong systems integrator partner/agency (shameless plug: Creatuity can be that partner if you need one) with composable commerce expertise, that opens possibilities. It's crucial to **match your platform strategy to your team's capabilities**. Also consider hiring/training: modern composable commerce might require frontend specialists, DevOps, solution architects – basically a different mix than a single-platform monolithic team. It's doable for mid-market (many have successfully made the leap), but you need to assess this honestly. A fantastic architecture on paper means nothing if you don't have the people to implement and run it smoothly.

Take a moment with these five questions. If you're scoring poorly on multiple fronts, it's a clear signal that sticking with the status quo will hurt your growth. The good news: you don't have to rip everything out tomorrow. There are smart ways to evolve your tech stack step by step – which leads us to migration paths to composable commerce.

Three Migration Paths to Composable Commerce

So you've decided you need to break out of the monolithic box. How do you actually do it without blowing up your business? From my experience, there are three pragmatic migration paths to consider. No one path is "right" for everyone – it depends on your pain points and priorities (hence why we started with those). Let's break down the options:

Path 1: Front-End First (Headless UI). This is the most common starting point I see, especially if your storefront experience is what's suffering. The idea is to **decouple your front-end from the monolith** and go "headless" – you keep your backend (product catalog, pricing, etc.) on the existing platform, but you build a new front-end experience that talks to it via APIs. Maybe it's a modern React PWA, a mobile app, or even a lighter website framework. The key is, the monolith's head (its built-in theme or front-end) is replaced with something faster and more flexible. For example, if you're on Magento or Adobe Commerce, you can swap out that dated Luma-based frontend for a fresh headless storefront. At Creatuity, we built a Headless Accelerator to do exactly this on Magento – basically an out-of-the-box modern front-end that can be implemented without replatforming. Solutions like this (or others like Vue Storefront, etc.) let you leap to a modern customer experience without touching your backend stability. The benefit? Users get a snappy, app-like site; you gain freedom to experiment with UX changes more rapidly. And importantly, your core commerce logic and data remain intact in the monolith for now, so risk is lower. I usually recommend front-end first **unless** your backend integrations are the bigger fire – in which case, see Path 2.

Path 2: Integration-Layer First (Composable Middleware). If your biggest headaches are around integrations (think ERP, CRM, PIM, or marketplace connections) or data consistency, attacking that layer first can yield huge wins. Many monolithic platforms encourage plugging integrations directly into their core (via apps, plugins, etc.), which as we discussed can be fragile. The composable approach is to implement an **integration hub or middleware** that sits between your monolith and the other systems. For instance, instead of a Magento extension for ERP sync, you might use an iPaaS (Integration Platform as a Service) like Celigo or MuleSoft, or even build a lightweight Node.js microservice that handles data flow. This hub pulls orders from the website and pushes them to ERP, and vice versa for inventory and product updates – all via APIs, not by gluing directly into the Magento code. The result? If the ERP side hiccups or needs to change, your eCommerce site doesn't crash because it's decoupled. One real example: a client had constant downtime due to their ERP integration extension failing. We replaced it with a cloud integration service (basically composable middleware), and suddenly updates could happen in near real-time without bringing down the site. Integration-first migration is great for **stability and data accuracy.** You'll likely reduce those "oversell" incidents where you sell stock you don't actually have (because inventory updates are faster and more reliable), and you eliminate that single point of failure from a bad plugin. The core site remains the same for customers, but behind the scenes you've created a more resilient engine.

Path 3: Greenfield Microservices (Sidecar or New Stack). This one's the boldest: you essentially start building a new mini-ecosystem alongside your monolith. I call it **sidecar microservices** – you identify a specific capability that your monolith does poorly (or not at all) and build it as a separate service that eventually could replace that part of the monolith. Some choose to do this as a "greenfield" project – e.g. develop a new microservice for order management, or a custom microservice for a unique fulfillment workflow – and run it in parallel with the old system until it's proven. For example, one of our clients needed far more flexible order management for omnichannel fulfillment than their platform offered. We helped carve that out into a composable OMS microservice running separately but hooked into the main site via APIs. It was built on modern tech (GraphQL, cloud functions, etc.) to handle complex order routing. Once live, if we needed to change something in the OMS logic, it was just a quick update to that microservice – no full-site regression testing needed. Another case: a client whose all-in-one system would crash whenever they ran heavy reports on big sales days – because the reporting jobs competed for resources with the live site. The sidecar approach was to build a separate reporting microservice with its own database optimized for analytics. That way, even if reporting got intense, it never threatened the checkout performance of the main site. Greenfield services are essentially the start of a new composable platform. It's like building the replacement car while still driving the old one – you move one piece at a time. The downside is it requires the most engineering effort up front, so I often see this when a company's tech team is ready for it or when the monolith is clearly at end-of-life and a full rebuild is on the horizon.

These paths aren't mutually exclusive. In fact, an ideal journey might involve all three in sequence: for instance, first put a headless front-end on (Path 1) to solve UX and site speed, then introduce an integration hub (Path 2) to solidify your data flows, then gradually replace core components with microservices (Path 3) over time. The key is incremental change. Each step should deliver a tangible improvement (faster site, fewer outages, etc.) and de-risk the next step. No "big bang" replatform where you disappear for 12 months and come back with an untested new system – that's how careers get ruined (and it's an anti-pattern we'll discuss shortly).

STACK OPTIONS: FROM SAFE BETS TO FULL MACH

Let's zoom out and talk about end-states – the stack options you might end up with after migrating bits of your monolith. Broadly, I see three archetypes of target architecture in the wild:

Monolith on PaaS (Modernized Monolith): This is essentially your original platform, but supercharged with better infrastructure and maybe modular tweaks. Instead of a cheap shared host or constrained SaaS, you run your monolithic platform on a scalable cloud platform (PaaS) or dedicated infrastructure tuned for it. You might containerize it, use auto-scaling groups, a CDN, etc., but you're not heavily breaking it apart. This option makes sense if your monolith actually meets most of your needs except scaling and performance. The baby apparel brand from our flash sale story eventually took this route: they moved off the overly constrained SaaS platform onto a PaaS where they could control server resources and database scaling. By splitting their read vs. write database traffic and adding a load balancer, they jumped from handling 5,000 simultaneous carts to over 12,000+ without crashing. They proved that with the right hosting and tuning, a monolith can stretch further. Monolith on PaaS is kind of the "safe bet" – fewer code changes, mostly infrastructure improvements. It gives you breathing room and buys time. But long term, you may still face the feature limitations and development slowdowns of the monolithic codebase. Think of it as extending the life of what you have while you plan bigger changes.

SaaS-Composable Hybrid: Here you blend the reliability of a SaaS platform with the flexibility of composable add-ons. Many mid-market companies go this route as a transitional state. For example, you might stick with a SaaS ecommerce engine (like BigCommerce, Shopify, etc. – whichever you're on) for core shopping functionality, but you replace or augment specific pieces with composable services. Maybe you plug in a third-party headless CMS for content, use a better search service, or build a custom microservice for subscriptions or B2B pricing – all integrated via APIs. The idea is to leverage the strength of SaaS (quick start, security, less maintenance) where it's "good enough", and overcome its weaknesses by bolting on better components where needed. A great example is Trade Tools, a 37-year-old industrial tools seller. They kept Magento/Adobe Commerce as the backend but went headless on the front-end and gradually swapped out components without customers even noticing. They introduced a new checkout and a PIM as separate services (classic composable move), resulting in a 31% jump in online revenue and 19% boost in conversion rate post-migration. All while lowering their total cost of ownership – they weren't paying for features they didn't need anymore and they could optimize hosting costs. That's the power of a thoughtful hybrid approach. Many retailers are also doing this by keeping, say, Shopify as the core but going headless for the frontend and adding custom functions via Shopify's APIs. The caution with a hybrid is to manage the complexity – you are mixing models, so make sure you have a strong integration and monitoring strategy. But done right, it's an excellent balance of stability and agility.

Full MACH (Pure Composable Microservices): This is the holy grail architecture we technologists get starry-eyed about – your entire stack is a collection of microservices, APIs, and headless front-ends, fully aligned with MACH principles. No traditional monolith at the core; instead, perhaps a set of cloud-native commerce services or a custom-built set of components. In practical terms, this might mean using a commerce platform that is headless and microservices-based out of the box (like Commercetools, MedusaJS, or others), or building a bespoke platform composed of many smaller SaaS and custom services orchestrated together. Very few mid-market companies start here, but some *do* reach this state after iterating through Paths 1-3 over a few years. What does full MACH get you? Maximum flexibility and theoretically unlimited scale. You can upgrade or replace any single component with minimal impact on the others. Your dev teams can work in parallel on different services, deploying continuously. It's the kind of setup Netflix or Amazon use internally – highly specialized services working in concert. One of our clients, a pure-play online retailer in the outdoor gear space, eventually ended up in a fully composable build: they use a headless commerce engine for the store-front, a separate PIM service, a custom-built checkout microservice, and an array of cloud functions for everything from image processing to fraud detection. This setup now runs like a well-oiled machine, because each piece is tuned for its job and scaling is as simple as turning up the dials on the services that need it. The trade-off: you now have many moving parts to manage. Without serious DevOps and architecture discipline, a full MACH stack can become overwhelming. But if you have the scale and team to justify it, it's a game-changer in terms of capability.

REAL-WORLD EXAMPLES: TRADE TOOLS, OUTDOOR GEAR, AND FARM & HOME

Let's cement these concepts with a few real-world case studies from my experience and podcast stories. Each shows different ways to balance monolith vs composable and the results achieved.

Trade Tools (Industrial B2B Hybrid): Mentioned above, Trade Tools is a decades-old B2B retailer in Australia selling industrial equipment. They had an old-school monolithic setup (Magento) that served both B2B and B2C customers. Instead of a risky full replatform, they took a hybrid composable route. They kept Magento as the system of record (products, orders, etc.) but implemented a headless front-end and began swapping out pieces modularly. One of the first swaps was the checkout: they replaced the native Magento checkout with a more optimized industry-specific microservice, reducing friction for users. They also integrated a dedicated PIM (Product Info Management) service to handle their large catalog better. These changes were done one at a time, carefully, so customers saw improved experience but no disruption. The impact was huge – they saw online revenue jump by 31% and conversion rate by 19% after launching the new headless experience. Better yet, their support tickets dropped because the site was more reliable post-composable migration. And since they weren't beholden to Magento's front-end or feature roadmap anymore, they lowered their long-term costs and future-proofed their tech stack.

Outdoor Gear Retailer (Mid-Market Pure-Play): This example comes from a client who sells adventure and camping gear entirely online (no physical stores). They loved their all-in-one SaaS platform initially for ease of use, but as they expanded their product lines and flash sales, they hit a wall on features and scalability. We took a phased composable approach. First, we implemented a headless CMS and a new front-end to give their site a fresh, dynamic feel (they wanted rich content and storytelling around their products which the monolith's template system struggled with). Next, we addressed checkout abandonment by plugging in a best-in-class checkout service that offered wallets like Apple Pay, etc., which their old platform didn't support well. Finally, we custom-built a microservice for inventory availability that could handle real-time demand during their seasonal sales (camping gear tends to spike in spring). This microservice was critical to prevent overselling hot items. Over 6 months, they basically morphed into a composable stack: the original platform became just one of the services (primarily for product catalog and order storage), with many new components surrounding it. The result: page load times improved by 40%, conversion went up significantly thanks to the smoother checkout, and they virtually eliminated oversell incidents (which used to plague them during sales). Importantly, they didn't do this all-or-nothing; at each stage they measured ROI – for instance, the checkout service paid for itself in three months via higher conversion. This outdoor gear retailer's journey shows that even a pure-play online brand can start with safe baby steps and gradually reach a modern architecture that keeps customers happy and the business nimble.

Farm & Home Omnichannel Retailer: I've worked on projects with multiple different large farm-and-home supply retailers (think rural lifestyle stores selling everything from pet feed to tractors). These are classic omnichannel businesses – big warehouses and stores plus a growing eCommerce channel. One such retailer was running on a monolithic platform with a bolt-on OMS and a clunky point-of-sale integration. The pain point was clear: **poor integration between online and store systems** – customers would buy online for store pickup, but inventory counts were often wrong, and store staff had to learn separate systems. We introduced composable components in a targeted way. First, we built a headless Order Management System that sat between the website and the stores. This OMS was custom-fit: it pulled online orders and routed them to the best fulfillment location (sometimes a store, sometimes a warehouse) based on inventory and proximity. It was tied into their existing admin UI, so store employees actually saw order info in the same interface they were used to (reducing training). Next, we deployed a **mobile POS** in stores that was integrated via APIs to the same system, enabling true buy-online-pickup-in-store (BOPIS) and even curbside pickup during the pandemic. Within months, they had capabilities like vendor drop-shipping and ship-to-store that their old vendor-locked system couldn't support. The compliance challenge (they sell firearms and ammo, which involve ATF regulations) was met by tailoring the composable components to handle those specifics – something no off-the-shelf software did out of the box. In the end, this farm & home retailer transformed into a poster child for omnichannel done right: customers got a seamless experience, and the company could rapidly roll out new fulfillment options (e.g., a 50-mile ship-from-store pilot to speed delivery) with confidence. They essentially turned their monolith-based system into a flexible hub by surrounding it with composable services. And when the next curveball came (like COVID-19 forcing curbside pickup), they were able to implement it in days, not months. That agility only comes when you're not shackled to a single rigid platform.

Each of these stories highlights a different angle – B2B hybrid, D2C pure-play, and omnichannel retail – but all share a common theme: pragmatism. None of these companies threw out everything and built NASA-level microservices overnight. They identified where the monolith was hurting them most, and they applied composable thinking to that problem first. Step by step, they evolved. And to me, that's the real playbook: *iterate your way to a better architecture*.

Beware the Anti-Patterns - What Not to Do

Just as important as what to do is what **not** to do. I've seen some approaches cause more harm than good – let's call out these anti-patterns so you can avoid the landmines:

Vendor Sprawl Without Strategy: It's easy to end up with 15 different vendors in your composable stack (each with monthly bills) and an ops nightmare. Don't replace one monolith with a *monster of mismatched services*. Composable isn't about using every shiny tool; it's about the right pieces. Avoid adding a vendor for every trivial feature

– consolidate where it makes sense. And always know who's responsible for what (there's nothing worse than an outage where 3 vendors all point fingers at each other). Keep an eye on total cost and integration overhead. If you find yourself needing a full-time person just to administrate SaaS accounts, reel it back in. Use best-of-breed, but be judicious and design an architecture where you could swap a piece out if needed - that keeps things flexible and vendors honest.

"Big Bang" Replatform Projects: I mentioned this earlier but it's worth reiterating: the big bang, where you attempt to rebuild *everything* and launch a completely new system in one go, is usually a mistake. It's high risk, high stress, and often fails to deliver because reality never matches the 12-month project plan. I've had companies come to me after a failed big replatform, practically in crisis mode. Users hate the new site, or it never launches, or it goes over budget by millions. Instead, gradually replace parts of the old system with the new until the old is gone. You get to deliver incremental value along the way and de-risk the transition. If an integrator or vendor tells you the only way to go composable is a full rebuild now, be skeptical. They might be chasing a big contract, not your best interests. I'll be candid: Don't do the overnight replatform if you can at all avoid it. There's a reason so many IT projects end in disaster. Evolve, don't flash-cut. Your sanity (and your customers' experience) will thank you.

DIY Everything: On the flip side, another anti-pattern is the "not-invented-here" syndrome – a tech team decides to build every piece from scratch in the name of composable purity. Listen, unless your business is building eCommerce software, you shouldn't be writing your own search engine, your own tax calculator, and your own cloud hosting platform. Leverage existing services and platforms where they make sense. Composable commerce has a vibrant ecosystem; you can buy or rent a lot of building blocks. Use your development muscle for what differentiates you. If ultra-fast search isn't your secret sauce, don't pour months into reinventing Elasticsearch. Focus on the areas that *do* set you apart – maybe that's a custom bundle builder for your products, or a pricing algorithm – and consider building those while integrating standard solutions for the rest. *Compose, don't code every-*

thing. The goal is a better outcome for your business, not just more tech for tech's sake.

In short, avoid the extremes of chaos (too many vendors, no plan) and over-ambition (rip-and-replace or overbuilding). Steer the middle course: deliberate, iterative, and focused on value.

ARCHETYPES AND PLAYBOOKS FOR COMPOSABLE COMMERCE

Every business is unique, but many fall into some common archetypes when it comes to how they should approach monolith vs composable. Here are three I see a lot, with their own playbooks:

B2B Sidecar Strategy: If you're a B2B company that's been primarily offline or on legacy systems (think big ERP driving the show), your path might be adding a sidecar digital channel. I call it sidecar because you're not overhauling the core system of record (often the ERP) immediately; instead, you bolt on an eCommerce component that rides alongside. The playbook here is to use composable principles to ensure this new sidecar doesn't turn into a mini-monolith. For example, many B2B firms launch an eCommerce portal that feeds off their ERP data but provides a modern front-end for customers. A **headless commerce platform** or custom frontend can sit on top of your ERP via APIs so that product, pricing, and inventory data flows from ERP, but the user experience is new and tailored for self-service buying. Ballard Industrial – a 70-year-old industrial supplier – did this by moving from just a static catalog site to a full self-service portal with headless commerce. They kept their ERP and back-office intact, but the new "sidecar" online portal handled customer-specific pricing, real-time inventory checks, and online ordering. It quickly led to an 83% increase in web traffic and won them industry accolades. The lesson for B2B: you often can't rip out your trusty ERP or back-office overnight, but you *can* deliver a composable online experience that interfaces with it. Prioritize integrations (so data is consistent) and build the front-end and new services in a modular way (so later you could even swap out that ERP if needed).

Omnichannel Retailer: This archetype covers any business juggling physical stores and digital. Your challenge is unifying customer experience across channels – store inventory visibility online, buy online/pickup in store, endless aisle, etc. The monolithic systems of old (enterprise ERPs, POS systems) often struggle with this cross-channel fluidity. The playbook: use composable tech as the glue and brain between channels. We saw this with the farm & home retailer example – a headless OMS to coordinate online and store operations, plus mobile POS to empower store staff. Another tactic is leveraging microservices for inventory and fulfillment logic: one service that knows all inventory across stores and warehouses and exposes that to any channel. This avoids each system (website, POS, etc.) having its own siloed view. An omnichannel retailer might still keep their existing POS in stores and an ecommerce platform online, but the composable layer connects them in real time. Also, consider composable solutions for in-store experiences – e.g., a tablet app for store associates that uses your eCom APIs to order out-of-stock items for customers (endless aisle). A lot of big-box retailers are moving this direction, but mid-market can too with today's cloud services. The key archetype principle: the customer should feel like they're interacting with one brand regardless of channel, so your tech's job is to compose a unified experience out of many systems. If your current architecture prevents that, inject a composable middleware or service to bridge the gaps.

Pure-Play Digital (D2C or Online-Only): If you're an online-only brand (no physical storefronts), you have one less channel to worry about – but you live and die by your online experience. The pure-play archetype is often where you see the most aggressive moves to headless and microservices, because the website is the business. The playbook here depends on scale: smaller D2C brands might actually start on a monolith (for speed) and only later sprinkle in composable elements as they grow, whereas a tech-driven startup might build composable from day one. Assuming you started monolith and are now feeling constrained (common case), focus on the elements that drive conversion and growth. That might be front-end flexibility for rich content (headless CMS, dynamic personalization), or it might be back-end scalability for peak events (microservices for cart/checkout as seen in our outdoor gear retailer example). Pure-plays often also need to experiment rapidly – A/B testing new features, launching micro-campaign sites, integrating the latest marketing tech – so a composable setup that allows quick additions and swaps is ideal. One pattern I see is using a core commerce engine (could be monolith or headless SaaS) and augmenting it with a bunch of Lambdas or cloud functions to run experiments or handle spikes. For instance, during a viral campaign, spin up a separate landing page app that feeds orders to the main system via API – if it blows up, your main site isn't overwhelmed. Pure digital players should measure tech success by how fast they can iterate (deploy frequency) and how well they can handle hyper growth without breaking (see: flash sale story). If your monolith slows you down on either axis, you know what to do.

Of course, some businesses might be a mix of these archetypes. The goal of defining them is to help you identify with a general strategy and learn from others who walked that path. Whichever archetype fits you, borrow the proven plays: sidecar approach for B2B, omnichannel glue for retail, speed and scalability tricks for pure-plays. There's no need to reinvent the wheel when you have a playbook in hand.

What to Measure: Building Your Composable Scorecard

"Improvement" is abstract unless you measure it. As you transition from monolith to a more composable engine, it's crucial to track key metrics that tell you if you're actually better off. I often help clients set up a simple "composable scorecard" dashboard. Here are some metrics that matter:

Deploy Frequency: How often are you deploying changes to production? This is a proxy for agility. If it used to take you a month to bundle changes for a release and now you're confidently deploying every week or even daily, that's a huge win. Higher deploy frequency (with solid quality) means you can respond faster to market needs. Track this number – it should go **up** as you embrace modular architecture and continuous integration. If it's not improving, find the bottleneck (maybe one of your "composed" pieces is still too tightly coupled or you need better automation). Remember, one hallmark of modern tech leaders is releasing *hundreds* of times a day – not saying you need that, but more often than you did as a monolith is a good target.

• **Oversell Rate:** This one is especially vital for retail/ecommerce – it's the percentage of orders that you later can't fulfill because the item was out of stock (i.e. you sold something you didn't actually have available). Oversells are a classic symptom of poor integration between inventory systems and the website. As you implement real-time stock updates via that integration hub or microservice, oversells should drop to near zero. Track it as a percentage of total orders or in absolute numbers. Every oversell is a disappointed customer and extra support cost. With composable architecture focusing on API-first inventory checks, the goal is to virtually eliminate oversells. If you still see them, that means some integration is lagging or not hooked up right – an opportunity for improvement.

• **Patch/Update Lead Time:** How long does it take to apply an important update or security patch to your system? In monolith land, even a minor version upgrade can be a project, because you have to make sure all parts (and extensions) still work. In a composable world, smaller services and SaaS components might update themselves or be easier to swap out. For example, if a critical vulnerability is found in your frontend framework, can you patch it and deploy in hours, or does it derail your whole week? Measure the time from "patch available" to "patch deployed". This should shrink considerably as your architecture modularizes. Faster patching means less security exposure and less time firefighting legacy issues. It's a good barometer of how nimble your tech operations are.

• **API Error Rate:** In a composable architecture, a lot of things communicate via APIs (internal or external). Keep an eye on the error rates of those API calls. This metric captures the reliability of your integrations. For instance, if your front-end is calling the product service and 0.5% of calls are failing, that could lead to missing info or broken pages. Or if your order service can't reach the tax calculator service 1% of the time, that's 1% of orders with potentially wrong tax (ouch). Use monitoring tools to track API success/failure rates and alert you to spikes. As you iterate, you want to see error rates go down or stay very low. Consistently high error rates mean something's flaky – maybe a service needs better scaling, or an API call needs a retry mechanism. Reducing this improves the customer experience (no more "Unable to load, please refresh" messages) and your team's sleep quality. In a monolith, these kinds of failures might have just crashed the whole app; at least with modular design, one failing API doesn't have to take everything out – but you still want it solid.

• **Page Speed / Uptime / Conversion:** Okay, sneaking in more than one metric here, but they are closely related to the above and to each other. Ultimately, you should track business metrics that tech improvements are supposed to drive. Page load times (especially on mobile) should improve with a headless or modern front-end – monitor Core Web Vitals or similar. Uptime (or conversely downtime incidents) should improve as you decouple pieces – one service failing shouldn't mean an entire outage. Count incidents and their severity. And conversion rate or revenue metrics are the acid test – did that 0.5s faster page or that new checkout service actually lead to more money? Tie your tech changes back to these KPIs. If you're not seeing any lift after major improvements, find out why and make adjustments. The beauty of composable architecture is you can tweak one part without undoing everything – so continuous improvement is part of the game.

Build a simple dashboard or report that your team and execs can see. It turns architectural evolution from a nebulous IT project into a measurable business initiative. It also keeps everyone honest – if switching to composable, track the metrics to ensure it's delivering the expected gains (and if not, learn and iterate).

90-Day Action Plan: From Idea to Implementation

By now you might be thinking, "This all sounds great, but how do I actually start?" Let's get super practical. Here's a step-by-step 90-day action plan I often recommend to clients ready to embrace a more composable approach. Consider this your checklist for the next three months:

Day 0–30: Assessment and Game Plan

1 Audit Your Pain Points and Systems: Gather your team (tech leads, ops, maybe a friendly customer service rep) and list out the current issues with your platform. Prioritize them. Is it site speed? Inventory sync? Inability to add new features? Be specific. Also inventory your current systems and integrations (what talks to what, where the data flows). This "state of the site" is crucial.

2 Define Your North-Star Goals: Decide what success looks like. Faster site by holiday season? Enabling BOPIS in 6 months? Cutting maintenance costs? Having clear goals will guide decisions and also help justify the project to any skeptics. Tie these goals to metrics from the scorecard we discussed (e.g. "improve conversion rate from X to Y" or "zero oversells next quarter").

3 Choose a Migration Path (or Two): Based on the pain points and goals, pick which of the three migration paths makes sense to tackle first. If front-end performance and flexibility is the issue, plan for a headless frontend project. If inventory and integrations are a mess, focus there first. This is where you outline your *strategy*: e.g., "We'll do front-end first, then tackle ERP integration in parallel." Keep it to one or two major initiatives to avoid overcommitting.

4 Rally the Troops: Communicate the plan to stakeholders. Get buy-in from execs by highlighting the business benefits (use those goals/metrics). Also, align your team – address any skills gap concerns (if you might need outside help or training for certain technologies, note it now). Set expectations that this will be an *iterative* rollout with continuous improvements, not a one-time big launch. It's important everyone is on board with the mindset.

Day 31–60: Quick Wins and Architecture

5. Implement a Quick Win Component: Identify something you can improve in a matter of weeks that demonstrates the value of composable architecture. Maybe it's deploying a new search service or a reviews widget via API, or standing up a small headless microservice for a niche feature. For instance, one client quickly integrated **ShipperHQ** for better shipping estimates as a composable add-on – a small win that built confidence. Quick wins create momentum.

6. Design the Target Architecture (Blueprint): While the quick win is being handled (perhaps by a subset of the team or an integration partner), have your architect or tech lead sketch the "blueprint" of how systems will interact after the migration. Define the major components (e.g., "React front-end talks to Magento via GraphQL API", "Orders flow from website to ERP via middleware X", "OMS microservice handles store fulfillment"). This doesn't have to be enterprise UML art – a simple diagram works. The point is to think through data flows early so you catch any big challenges (like "oh, our ERP API can only handle 5 calls/minute, we might need an upgrade or a staging DB").

Also decide on integration patterns – will you use a message queue, webhooks, bulk data syncs, etc. for each connection. Basically, **measure twice, cut once**: get the architecture clear on paper now.

7. **Choose Your Tech Stack & Partners:** Select any new platforms or services you'll need. If going headless front-end, evaluate the options (build your own vs. use a framework like Next.js or an accelerator like Creatuity's). If using an iPaaS for integration, pick one now. Also choose implementation partners if needed – maybe you'll engage an agency for the headless build or a consultant for DevOps. Lock those in during this window so that execution can start. Ensure any chosen tech aligns with your team's skill or there's a plan to bridge the gap. For example, if none of your devs know React, maybe part of the plan is hiring or contracting a React expert. It's better to address that now than during crunch time.

8. **Data and Dependency Check:** Do a sanity check on your data. Are there any data silos that need cleaning up before integrating? For instance, is your product data all in your eCommerce DB or split with an ERP? You might find you need a mini project to consolidate or clean data (like making sure all products have unique SKUs across systems). Also list out any dependencies – e.g., "We can't deploy new front-end until the API is upgraded to version X" or "We need SSL certs and cloud environments set up for the new services." Work on these foundational tasks now so they don't become blockers later.

Day 61–90: Execution and Launch

9. **Build/Integrate in Sprints:** With architecture and tools decided, get to building the first major piece of your composable puzzle. Use agile sprints (2 weeks is common) to develop and integrate. Whether it's the headless front-end or the new middleware, break the work into increments. Have demos at the end of each sprint to show progress to stakeholders – this keeps everyone excited and spots issues early. Also plan your data migration or syncing steps if any (for example, if moving to a new PIM, make sure product data is being loaded properly).

10. **Test the Heck Out of It:** As you approach the end of the 90 days, focus on quality. If it's a front-end, run user experience tests and performance tests (remember, one goal is likely a faster site – measure

it). If it's an integration, do test orders, simulate inventory changes, etc., to ensure things flow. The good news: because you're adding components rather than replacing everything at once, you can often test in parallel with the live system. For instance, we could run the new headless site on a staging URL with real data to see how it behaves under load *before* swapping it in. Do not skimp on testing – the confidence to deploy comes from here.

11. **Gradual Rollout:** Plan a phased launch if possible. Maybe release the new front-end to a small percentage of users (feature flags or a beta program) and monitor metrics. Or roll out the new integration for one category of products or one region first. This way, if something goes off, you can roll back or fix forward quickly without a full-scale disaster. Many composable architectures allow toggling components on/off because of their modular nature – leverage that.

12. **Go Live and Monitor:** Flip the switch – push the new component live for all users. Congrats! But your job isn't done. Closely monitor those key metrics from the scorecard. Is load time indeed better? Any error spikes? Keep an eye on logs, set up alerts on KPIs (like API errors or spike in cart abandonment). Have your team on standby to address any post-launch issues immediately. Usually, with good prep, things will run smoothly, but monitoring ensures you catch anything unexpected (maybe an edge-case integration bug) before it becomes serious. Over the next few weeks, gather feedback from customers and staff – any complaints or praises? Use this to iterate.

By Day 90, you should have at least one composable component or improvement in production, delivering value. Maybe it's modest – that's okay. The important part is you've proven you can *do it*. From here, you'll rinse and repeat: tackle the next pain point with the same assess → plan → implement cycle. Each iteration moves you closer to that ideal scalable engine.

What's Next: Data Plumbing and Future Growth

Congratulations – if you've followed along, you've learned how to rethink your ecommerce platform and take the first steps toward a composable, scalable architecture. We took a journey from a flash sale horror story to a practical action plan, covering the why, what, and how of monoliths vs composable systems. By now, you should have a solid framework to evaluate your own stack and a playbook to start evolving it.

The next chapter is all about making sure the **pipes** connecting your shiny new composable components are clean and efficient. We'll dive into *data plumbing* – syncing your ERP, PIM, inventory management, and other enterprise systems into this new architecture. After all, even the sleekest microservice frontend means nothing if it's showing wrong stock or outdated product info. We'll talk integrations, data models, and the often unsexy (but critical) work of ensuring information flows correctly through your commerce engine. Think of it as hooking up the plumbing in the dream house you're building.

3.3: DATA PLUMBING – ERP, PIM & REAL-TIME INVENTORY

A few years back, I got an early-morning call from a panicked B2B merchant who sells industrial pipe fittings. They had just oversold a critical part for a construction project. Here's what happened: a customer ordered 200 specific fittings for next-day delivery to a job site. The e-commerce site showed 200 in stock, so the customer paid and went to bed expecting their order. But unbeknownst to anyone, *another* order for 20 of those same fittings had been placed hours earlier – and that order was stuck in a delayed batch sync between the site and the ERP. The warehouse only had 180 pieces on the shelf. Cue a frantic scramble: the team discovered the shortfall too late, the construction crew sat idle, and the merchant ended up chartering an emergency shipment from another distributor. All told, that data disconnect **cost over $200,000** in rush fees and penalties. One little sync delay, one big nightmare.

Unfortunately, overselling fiascos like this aren't one-off flukes – and the stakes are higher than ever. We live in an age of real-time expectations, yet I still meet e-commerce teams dragging CSV files around manually to update inventory or pricing. If you're downloading a spreadsheet of stock levels from your ERP each night and uploading it to your site the next morning, you're playing with fire. A lot can happen in those 24 hours – and each oversell or pricing mistake eats directly into your margin. Do the math: if you oversell even five orders a week and each incident costs you a $500 appeasement or lost future business, that's $130,000 vaporized in a year. The cost isn't just financial, either. It's angry customers, bad reviews, and brand damage.

Beyond the immediate dollars and cents, e-commerce data plumbing has become a make-or-break issue because of how fast the landscape is evolving. This isn't 2015 anymore – your data infrastructure needs to serve many stakeholders. It's not just your website and your warehouse in the loop now; it's also marketing platforms, AI services, marketplaces, maybe even voice assistants. One huge new factor is what I call *AI Visibility*: if your product data isn't clean, structured, and accessible, you risk being *invisible* in the era of ChatGPT and AI-driven shopping. Imagine a consumer saying, *"Hey ChatGPT, what's the best stainless steel pipe fitting I can get by tomorrow?"* If ChatGPT doesn't know your products exist, they'll never get recommended. Feeding timely, machine-readable data to AI channels is the new SEO – and if you're still struggling just to keep basic inventory in sync, you're not ready for that world.

So, what exactly is "data plumbing" in e-commerce? It's the **pipes and pumps that move your critical data** – products, inventory, orders, customer info, prices – between all the systems that need it. Done right, it's invisible, flowing in real time. Done wrong, it's a leaky mess of manual workarounds and nightly cron jobs waiting to fail. Let's talk about how to fix the leaks and build a modern data plumbing stack that can handle the growth.

Mapping Your Modern Data Pipeline

In a modern commerce stack, data typically flows through a chain of connected systems – think of it like a series of tanks and pipes. It often starts with your **ERP (Enterprise Resource Planning)** system, where things like inventory counts, product SKUs, and maybe accounting or pricing live. Many businesses treat the ERP as the "source of truth" for inventory and orders. Next, you might have a **PIM (Product Information Management)** system dedicated to product content – all the marketing descriptions, images, specs, and attributes that don't belong in a financial system but are gold for selling online.

The PIM often pulls base product data from the ERP (like SKU codes or initial specs), enriches it, and then feeds it to your **e-commerce platform** (whether that's Shopify, Magento/Adobe Commerce, BigCommerce, etc.). The e-commerce platform is your online storefront – it displays the product info, takes the orders, and usually manages the shopping cart and checkout.

From there, orders might flow out to an **OMS (Order Management System)** or back to the ERP for fulfillment. Meanwhile, inventory levels need to flow back from the warehouse (via the ERP or a warehouse management system) to the e-commerce frontend to update what's available to sell. Customer data and behaviors (browsing activity, purchase history, etc.) may flow into a **CDP (Customer Data Platform)**, which aggregates everything into unified customer profiles for marketing and personalization. And don't forget the myriad **plugins and integrations** on the periphery: email marketing tools, analytics dashboards, review platforms, ChatGPT plugins – all of them need hooks into your data pipeline too.

It sounds like a lot – and it is – but mapping it out helps. In an ideal setup, each system in this chain talks to the next *automatically and instantly*. For example, when inventory is received in the warehouse and logged in the ERP, the ERP should immediately push an update to the PIM/e-commerce platform so the website shows the new stock. When a new product is created in the PIM, it should sync to the site with all its content. When an order is placed on the site, it should shoot over to the ERP or OMS within seconds for fulfillment. And all along, your CDP is following along – ingesting stock levels, product info, and customer interactions in real time so it can drive things like personalized recommendations or trigger a back-in-stock email.

That's the ideal. But many mid-market merchants I meet have a very different reality: siloed systems held together with duct tape (or weekly CSV uploads). How do you get from point A to point B? It helps to take a phased approach. I often advise a **crawl-walk-run** framework for scaling up your integrations.

CRAWL, WALK, RUN: SCALING YOUR INTEGRATIONS

When you're just starting out or still small, you can *crawl* – do things in a simple, maybe semi-manual way – and get away with it.

But as you grow (walking, then running), your approach to integration needs to advance. Here's how I break it down:

Crawl – Basic & Manual: In the crawl stage, you might have minimal automation. Maybe you're updating inventory once a day by exporting from your ERP and importing to your site, or you're manually keying online orders into your accounting software. This stage is okay for a brand-new eCom operation or a side project where volume is low. The focus here is just getting data moved *somehow*. It's not pretty (and definitely not real-time), but it works in the short term because the scale is manageable. The key is to **document** these processes and start logging pain points – e.g. "it takes 3 hours every Monday to merge spreadsheets" or "we keep selling items that sold out over the weekend."

Walk – Automated & Batch: At the walk stage, you invest in **automation for the high-priority data flows**. Instead of manual CSV transfers, you set up scheduled jobs or use an integration app. For example, you might configure your e-commerce platform to fetch inventory updates from the ERP every hour, or use a connector service to push orders into the ERP in batches every 15 minutes. Latency is improving – maybe your inventory is at most 15 minutes out-of-date instead of 24 hours. This is where many growing merchants land: using an iPaaS (Integration Platform as a Service) or built-in platform connectors to automate regular syncs. It's not fully real-time, but it's a lot better than manual or once daily. The goal in this phase is to eliminate human swivel-chair tasks and reduce the window for errors. If you're at this stage, you should also start implementing basic **monitoring** – even if it's just an email alert if a scheduled job fails or if an import doesn't balance out the stock numbers.

Run – Real-Time & Composable: Now we're sprinting. At run stage, you aim for **near real-time, event-driven integration**. When something happens in one system, it's pushed immediately to the others via APIs, webhooks, or message queues. Inventory changes propagate in seconds. Orders flow instantly. Price updates go live without delay. Achieving this often means building more **modular, composable connectors** rather than relying on inflexible off-the-shelf plugins. You might introduce a lightweight middleware layer – essentially a translation and routing service that sits between, say, the ERP and your eCommerce platform, making sure they speak the same language. This stage also brings sophisticated monitoring and logging; you have dashboards showing the health of each pipeline (e.g., "Inventory sync: 98% of updates last 5 min, 2% took longer than 1 minute"). The run stage is where you truly minimize latency and manual intervention, but it comes with engineering investment. The payoff is **resilience and agility** – your systems can handle spikes and you can add or swap out components (like an ERP or storefront) with less pain because everything is decoupled.

Not every business needs to sprint at full speed in every area. Part of this crawl-walk-run model is deciding **where** real-time really matters. For instance, maybe real-time inventory is critical (to avoid those oversells), but you can live with batching product info updates once a day. Or vice versa: your product content might update constantly (think price changes in a dynamic pricing scenario) while inventory is steady. Define what needs to be real-time versus periodic based on cost/benefit. This ties directly into the next point: planning your integration with clear goals.

ERP INTEGRATION PRINCIPLES: FROM "DO WE NEED THIS?" TO OBSERVABILITY

Let's start with the elephant in the room: **your ERP itself**. One of the first questions I often ask a client struggling with integration is, *"Do we even need this ERP connected (or at all)?"*. Don't get me wrong – ERPs can be fantastic for managing complex operations. But I've seen cases where a business is using an ERP that's complete overkill (maybe adopted too early or forced on them by corporate), and it actually creates more problems than it solves. An overly complex or ill-suited ERP will make integration a nightmare – endless custom fields that don't match your eCom platform, brittle batch processes, etc. So step one, **audit your ERP**: is it the right fit? Is every piece of data you're syncing truly needed, or are you contorting your e-commerce around an old piece of software? I've literally recommended pausing an eCom project to switch ERPs or even drop the ERP entirely in favor of simpler systems for a client's current stage. If the ERP is non-negotiable (I know many CFOs clutch theirs dearly!), then we commit to making it work better.

Assuming you're sticking with your ERP, the integration needs a **clear goal line**. Define what success looks like before you start wiring things together. For example: *Do we need real-time inventory sync or is 15-minute sync okay? Do we want **every single field** automated and integrated or just the 80% that matter?* Setting these parameters guides the entire project. I always emphasize the **80/20 rule**: automate the 20% of processes that consume 80% of your team's time, and don't obsess about the edge cases upfront. If 80% of your orders are straightforward but 20% require some special handling, you might integrate the straightforward ones first and handle the rest manually until you can iterate.

Next, **architect the integration for flexibility and minimal invasiveness**. Ten years ago, the "state of the art" was buying a giant monolithic extension or plugin that you install directly on your e-commerce platform to talk to your ERP. I did plenty of those in Magento back in the day – you'd buy, say, an "SAP integration module" and drop it in. Those tended to be huge, black-box bundles of code (often encrypted) that attempted to sync *everything*, and they were notoriously brittle. If the ERP updated or your platform updated, things broke. If you didn't use a particular ERP feature that the module expected, you had to live with errors or extraneous data sync. In 2025, I strongly recommend a **composable, modular approach** instead. Build or leverage smaller components: one service or function for inventory sync, another for order push, another for product data, etc. For example, at my agency Creatuity we integrated an *ancient* ERP called Epicor Eagle for a retailer (this ERP predates the internet era – it was spitting out fixed-width text files!). We ended up writing a set of connectors that translated Eagle's funky file format into modern API calls to Magento. We didn't touch the Magento core at all – no plugin installed there. The beauty is, when that retailer later considered switching their storefront from Magento to Shopify, our integration work wasn't lost – we could reuse the ERP side connectors and just swap the eCom side to talk to Shopify's API. **That's the power of composable plumbing**: your pipes have joints, so you can re-route them without ripping every-thing out.

Another best practice: **use an iPaaS if it fits**. Integration Platform as a Service tools (like iPaaS.com, Celigo, Boomi, MuleSoft, and others) provide pre-built connectors for common systems. If you find one that supports both your ERP and your eCom platform out-of-the-box, it can jump-start your project. Why custom-build everything if someone's already paved that road? The caveat is to ensure it meets your needs (latency, customization, cost) and doesn't just become another "black box." Sometimes an iPaaS can handle the "walk" stage of integration for you, and then if you outgrow it, you might graduate to a fully custom "run" solution later. And sometimes, the iPaaS is robust enough to stick with long term.

No matter how you integrate, **data consistency is key**. Watch out for format mismatches and hidden assumptions. One client of ours discovered their ERP stored quantities as whole numbers only (no decimals) while their eCom needed to handle 1.5 units (for selling half quantities of a material). The first time 1.5 went from platform to ERP, it silently became 1 – talk about an inventory nightmare. The lesson: **standardize your data formats** and units early. If one system calls something "XL" and another calls it "Extra Large", map or reconcile that. If one uses UTC timestamps and another local time, fix it. These little inconsistencies will bite hard at scale.

Perhaps my biggest piece of advice for ERP integration (or any crit-ical integration) is to **embed observability and alerts from day one**. Don't treat integration as a "set it and forget it" thing. Assume that delays and failures *will* happen, and instrument accordingly. That means logging every sync attempt, timing how long it takes, catching errors, and sending up a flare if something goes wrong. For instance, if your goal is real-time inventory, build a monitor that flags if any inven-tory update has taken more than say 5 minutes to appear on the site. If an order fails to push into the ERP, make sure an email or Slack message goes to the right person *immediately*. Trust me, you do **not** want to find out about an integration issue only when customer service says "Hey, customer John Doe is asking why his order from 3 days ago hasn't shipped – turns out it never hit our ERP." Use logging and alerting tools – even simple ones – to keep eyes on the pipes. Modern observability stacks can get fancy (tracing, dashboards, etc.), but at minimum have error notifications and daily reports of sync status.

To sum up ERP integration: **audit the need, set clear goals, go modular, ensure data consistency, and make it observable.** If you do all that, you'll avoid the worst pain points. And one more thing – don't be afraid to leverage **AI in the integration process itself.** This is forward-looking, but we've started adding small AI-driven components in our data flows. For example, one client didn't have a PIM and was pulling product descriptions straight from ERP (which had bland ones). We inserted an AI enrichment step – basically, as product data flows out of the ERP, an AI (via an API call) adds a bit of marketing copy or flags if the description might violate some compliance rule (like for regulated products). This kind of inline AI can improve data quality on the fly. Because our integration was composable, adding that AI piece was as simple as plugging in another connector.

PIM: The Single Source of Product Truth (and Five Big Wins)

If you don't have a PIM yet, you've probably felt the pain of managing product data in a patchwork of spreadsheets and admin panels. I've seen **spreadsheet chaos** up close: one client had different Excel files for specs, translations, pricing, and images, spread across various departments. Launching a new SKU was like herding cats – copy-pasting info from sheet to sheet, emailing for missing data, often ending in errors or missing info on the live site. In fact, it got so bad they sometimes just gave up on adding new products because it was too time-consuming. That's death for growth. We helped them implement a PIM and the results were dramatic: time-to-market for new products shrank by weeks, and they were able to push 30% more new SKUs live each quarter. The PIM centralized all the product workflows and stopped the bleeding.

So what *is* a PIM? It stands for **Product Information Management** – and it does exactly what it sounds like: helps you manage your product data in one place. I often describe it as "the CRM for your products." Just as a CRM forces salespeople to fill in all the details about a lead, a good PIM forces your team (or your suppliers) to fill in all the details about a product. **The power of a PIM is being** *the* **source of truth for product content in a multi-channel world**. Here are five big wins you get from deploying a PIM:

Data Centralization: All product info – names, descriptions, specs, dimensions, images, PDFs, compatibility info, you name it – lives in one structured repository (not scattered across files or systems). This single source of truth means when something changes, you update it once and it propagates everywhere. No more inconsistencies where your website says one thing and your catalog PDF says another.

Improved Data Quality & Completeness: PIMs let you enforce rules and workflows. You can require that every product has a weight, or that every "smartphone" category product has a value for "screen size" attribute, etc. It's like a digital checklist that won't let a product go live until it's complete. This prevents those embarrassing "Image coming soon" or "Lorem ipsum" placeholder text moments on your site. High-quality, detailed product info boosts sales and reduces returns – customers aren't going to buy if your product page has just a SKU and no info. Conversely, rich descriptions and specs increase confidence (we actually saw a 25% conversion rate lift for a grocery retailer who added detailed info and images via a PIM, after originally having bare-bones listings).

Faster Time-to-Market: As illustrated with the spreadsheet chaos example, a PIM streamlines the process of getting new products online. Multiple team members (or even vendors) can collaborate in the PIM concurrently – one person writing descriptions, another uploading photos, another adding technical specs. You can have workflows for approvals. This parallel processing slashes the delay. That mid-market brand I mentioned went from weeks of lead time to launch a product to just days after adopting a PIM. Speed here means competitive advantage, especially if you have seasonal product lines or need to react quickly to trends.

Multi-Channel Consistency: In 2025, you're likely not just selling on your own website. You might be on Amazon, eBay, Google Shopping, physical stores, maybe even feeding data to a mobile app or voice assistant. A PIM can syndicate and format your product data to all these channels from one hub. Change a price or description once, and push it everywhere. It often supports custom export templates for each channel's format. Some PIMs even integrate with marketplaces directly to publish products. The result is consistency – customers get the same accurate info whether they see your product on Instagram or on your own site. (And if you need to pull a product for compliance reasons, you do it once in the PIM and it's gone from all channels.)

Scalability & Governance: As your catalog grows (hundreds, then thousands, then tens of thousands of SKUs), a spreadsheet approach falls apart. PIMs are built for scale – robust databases that can handle lots of records and relationships (like products with variants, bundles, etc.). They also provide audit logs, versioning, and role-based access, which is important so you know *who* changed the diameter of that pipe fitting from 5cm to 6cm and when. Treating your product data with the same rigor as, say, financial data, pays off in fewer mistakes and easier onboarding of new team members (the process is standardized).

When selecting a PIM, you have options ranging from commercial SaaS platforms to open-source solutions. Some of the **leading PIM vendors** in the market include Salsify, Akeneo, InRiver, Sales Layer, and Informatica (for big enterprises). There are also **open-source PIMs** and suites: for example, **Pimcore** is a popular open-source platform that includes PIM and other modules; there's also an open-source project called **OpenPIM**, and others like **Ergonode** or **AtroPIM** that cater to different needs. Akeneo offers a Community Edition that's open-source (and a paid Enterprise Edition with more features). The trade-off generally is cost vs. flexibility: a commercial PIM (like Salsify or Akeneo Enterprise) will come with support, perhaps a faster implementation timeline, and a polished UI – but you pay license fees or subscriptions. An open-source PIM might save licensing costs and give you full control (free as in free speech, if not as in free beer), but you'll need the technical talent to host, customize, and maintain it. **No one-size-fits-all** here – match the choice to your budget, in-house skills, and integration needs. Whatever you choose, plan for it to be a long-term investment; you ideally won't rip and replace your PIM often (far less frequently than, say, redesigning your website).

Now, let's talk about a fun new frontier: **AI enrichment for product data**. We now have AI tools that can significantly ease the burden of creating and maintaining great product content. I've seen brands do clever things like export their entire product catalog to a spreadsheet, feed it into ChatGPT, and have AI generate improved descriptions, SEO keywords, even preliminary category assignments – then have humans review and import the results back into the PIM. One retailer presented about this at a Meet Magento event and showed how they slashed their manual data cleanup time while accelerating product launches. It basically turned a month of writing work into a few days of AI + editing. I'm also excited about integrating AI *directly* into the PIM workflows. Imagine as soon as you create a new product, an AI agent auto-fills a draft product description, suggests three upsell items based on similar products, and flags if the description might have compliance issues (like mentioning a banned term). This isn't sci-fi – with tools like OpenAI's APIs or frameworks like LangChain, you can start embedding these capabilities into your product pipeline today. In fact, if you glance at the resource list at the end of this book, you'll find a link to a **PIM AI Enrichment Checklist** that outlines steps to get started. The bottom line: a PIM sets the stage for consistent, high-quality product data, and layering AI on top can unlock even more efficiency and creativity.

Before we move on, let me hammer this home: *If you don't have a PIM, seriously consider getting one.* Your sanity (and your team's) will thank you, and your customers will have a much better experience. Now, with products and inventory flowing nicely, let's address how to **keep that data accurate** – especially inventory and pricing – so we don't get any nasty surprises like our pipe-fitting saga or pricing glitches.

REAL-TIME INVENTORY AND PRICING SAFEGUARDS

Nothing will destroy customer trust faster than selling something you can't actually deliver. We opened this chapter with a horror story of overselling due to slow inventory sync. To avoid that, **real-time (or near-real-time) inventory updates** are the gold standard. If at all possible, your systems should update inventory *immediately* when sales happen or new stock arrives. Many e-commerce platforms have features like "incremental inventory adjustments" (e.g., when an order comes in, it can decrement stock right away in the database, even before the ERP confirms). Use those features! Also, consider a bit of **safety stock buffer** – for example, show 98 available when you actually have 100, just to cover any small discrepancies or damaged items. But buffers only get you so far; the true fix is rapid, accurate syncing.

If real-time isn't feasible across the board (maybe your ERP is too slow to handle that load), at least identify your hottest-selling or most critical products and make sure their stock is updated frequently. I've seen retailers do 5-minute interval updates for their top 100 products, and daily for the rest. Not perfect, but better than nothing. And always have a plan for **what to do if an oversell does happen**: for instance, automatically send an apology email with a coupon if an item sold out, or have an "oversell reserve fund" in your accounting so you acknowledge it as a cost of doing business (which motivates you to minimize it).

Now onto pricing – the other potential disaster zone. We've got to talk about the infamous Cartier glitch. This is one of my favorite cautionary tales. Cartier's US e-commerce site once had a currency conversion error where they defined USD as base but mis-set the conversion for Mexican pesos. They basically dropped three zeros – so an item that was 237,000 pesos (about $14,000) got shown as 237 pesos, about $14. A customer in Mexico managed to buy two Cartier LOVE earrings – which are $14k – for equivalent of $14. And here's the best part (or worst part if you're Cartier's ecommerce team): Mexican consumer law forced Cartier to honor that price, so they had to deliver those $14,000 earrings for pennies on the dollar. The customer even refused Cartier's desperate offer of champagne and perks to cancel – they wanted their bargain bling. Ouch.

How did this happen? Essentially, a *single API misconfiguration* and lack of failsafes. The site's currency conversion API for MXN was faulty and there were no checks in place. Similarly, I've seen a case where a shipping rate API went down, and because the site didn't have a backup or error handling, it defaulted to $0 shipping – customers suddenly got free overnight shipping until someone noticed. These scenarios can cost millions in a day if you're not careful. And it's not theoretical; it *will* happen if you run enough transactions for long enough without safeguards.

So what do we do? We implement **guardrails and monitors** – essentially, a pricing and inventory safety checklist. Here are some key safeguards I insist on:

Audit and Failover Plan for Every Critical API: Identify every external or internal service that affects price, inventory, or shipping costs – currency converters, tax calculators, shipping rate APIs, inventory feeds – and make sure there's a **failover strategy**. That could mean a secondary provider (e.g. use a backup currency conversion API if the primary fails, or have flat-rate shipping kick in if FedEx's API is down). It could also mean a cached last-known value that you use temporarily. Map these out and document them. If something as important as currency conversion is single-threaded with no backup, you're one glitch away from disaster.

Threshold Alerts for Anomalies: Establish rules for what "unusual" looks like in your system and trigger alerts or automated actions. For example, if a product's price suddenly drops by more than 50% in a single update, flag it before it goes live. If an order comes in with total $0 (when it shouldn't), that's a red flag. If an order is 10x your average order value, maybe hold it for review (could be a pricing error or fraud). These can be simple rules – you **don't** need an AI PhD for this. As I often quip, *rule-based systems still work; you don't need AI to flag orders way above or below average*. A few "sanity check" if-then rules can catch the obvious oopsies. Modern e-com platforms or monitoring tools let you script these or use built-in features (for instance, some platforms let you set a max discount or min price for products).

Real-Time Error Monitoring & Alerts: Use tools like New Relic, Sentry, or simple log monitors to watch for exceptions or unusual events in real time. If your inventory sync script throws an error at 3 AM, someone should know at 3:01 AM, not at 10 AM when customer service starts getting calls. Set up alerting pipelines (PagerDuty, Slack, email – whatever your team will see immediately). And importantly, **assign responsibility**: who gets the alert and has the authority to act? If a price glitch is detected, is there an on-call developer who can yank that product off the site or do you empower an e-commerce manager to disable checkout for that SKU? Define that process in advance. In Cartier's case, if someone had been alerted the moment the first $14 order came in, maybe they could have paused MXN currency sales and saved themselves a fortune.

Human in the Loop for High-Stakes Changes: Despite all the automation, I'm a fan of a human check for certain things. For instance, if you're doing a bulk price update affecting thousands of products, maybe have a dry run output a report of the changes and have a person glance at ranges ("are we sure these ten items should really be $0 or 90% off?").

Regular Drills and Performance Tests: Chaos engineering isn't just for big tech companies. You can do mini fire-drills: "What if our inventory feed stops? What if our pricing service returns null?" Have a playbook for those scenarios. Also, test your integrations under load – if you have a big sale event, ensure your pricing API can handle the volume, or have a fallback (e.g., switch to static pricing if the dynamic service lags). If your systems are slow and orders queue up, that's when mistakes slip through. Performance test and set SLAs for critical services.

Implementing these safeguards might sound like extra work, but one $14 Cartier earring incident covers the ROI of a whole lot of preventative effort. The **Resources** section has a "Pricing & Inventory Fail-safe Checklist" – I highly recommend going through it line by line with your tech team. It's far better than going through a post-mortem after losing six figures to a glitch.

UNIFIED CUSTOMER DATA & AI VISIBILITY: CDP TO THE RESCUE

We've talked about product and order data, but what about *customer* data? In e-commerce, understanding and leveraging customer information can be as important as the products you sell. This is where a **Customer Data Platform (CDP)** comes into play as part of your data plumbing. I like to think of a CDP as the brain that all the other systems feed into and draw insight from. It's similar to a CRM (Customer Relationship Management) system, but broader in scope. **A CDP is exactly what it sounds like – a platform designed entirely for holding and unifying your customer data.** Unlike a CRM which might just have sales or support info, a CDP pulls in **everything**: on-site browsing behavior, email engagement, in-store purchases, customer service interactions, loyalty status, third-party data, etc. The goal is a

unified profile of each customer that any tool (marketing, support, personalization engine) can tap into in real time.

Why is this a must-have? Because customer expectations for personalization are sky-high, and because AI-driven experiences (like recommendation engines or chatbots) are only as good as the data you feed them. If your data plumbing for customer info is poor, you get things like sending a "We miss you, here's 10% off" coupon to someone who *literally bought something yesterday* under a different email address. Or showing the same product recommendations to a repeat customer that they already bought last month because your systems didn't talk to each other.

A CDP helps avoid those blunders by letting you segment and trigger based on the *complete picture* of a customer. For example, with a CDP you can do things like: "Customers who browsed Category X on the website and purchased in-store in the last 30 days, send them this specific follow-up offer via email." You can't do that reliably if store and web data aren't merged. Home Depot does this well: if you use your loyalty account in-store, their CDP is syncing that behavior and might text you an offer before you even exit the aisle. That's the power of connected data – online and offline blending into one experience.

If you're considering a CDP, some of the top commercial names out there are **Twilio Segment, Adobe Real-Time CDP, Salesforce CDP, Tealium, and Treasure Data**. For mid-market companies, Segment has been popular (easy to implement, good integrations), though note that costs can rise as your data volume grows – at a very large scale, some find Segment gets pricey. Adobe and Salesforce's CDPs make sense if you're already deep in those ecosystems (e.g., using Adobe Commerce/Magento or Salesforce Commerce Cloud – they'll naturally tie in). Tealium and Treasure Data are also powerful, with Tealium often known for its real-time capabilities. On the open-source/affordable end, there's **RudderStack** (which is essentially an open-core Segment alternative, where you can self-host some components), and interestingly **Pimcore** has a CDP module now, which plays nicely if you're using Pimcore for PIM as well. Apache Unomi is another open-source CDP geared toward developers comfortable with the Apache stack.

Adopting a CDP is not a trivial project – it requires integration work (feeding data from all sources in, and then pushing segments out to your marketing tools), and careful consideration of privacy/GDPR compliance (a good CDP will have built-in tools for "right to forget" and consent management). But the payoff is huge. I had an e-commerce exec tell me, **"I wouldn't launch a new e-commerce site without implementing a CDP first"**. That's how foundational it's becoming to unify data from day one.

From a data plumbing perspective, think of the CDP as both a **repository and a distributor**. It sucks in data from your e-commerce platform, your ERP (customer purchase history), your email system, POS, etc., and then it can also feed that enriched profile data back out to where it's needed. For example, your website personalization engine can call the CDP in real-time to decide what banner to show ("Oh, this user is a high-value repeat customer, show the VIP promo"). Or, very timely for 2025, your **ChatGPT plugin or AI assistant** could query the CDP to give a personalized response. Picture a customer asking a chat-bot, *"Hey, do I have any rewards points I can use right now?"* – with a CDP, the bot can actually answer that, because it has a unified profile with loyalty points and purchase history.

This brings us to **AI Visibility Enhancement** – a concept I'm really passionate about. We touched on it earlier: making your data available and optimized for AI consumption. In practical terms, this means two things: (1) preparing your product data for AI-driven discovery (like ChatGPT plugins), and (2) using AI to act on unified customer data for personalization. The first is more about being visible in new channels, the second is about improving your own user experience.

To **enhance AI visibility** for your products, you should ensure your data plumbing allows for easy access to your catalog by AI services. If you're on Shopify, for instance, they already have a public ChatGPT plugin that can surface Shopify store products (so if you opt in, ChatGPT can pull your product info when users ask). There's also the Klarna plugin which pulls in product data across many stores to answer shopping queries. Make sure you're participating in these where it makes sense. If you're a larger brand or want more control, you can even **build your own ChatGPT plugin** for your store – essentially an API with an OpenAPI spec that ChatGPT can call to get your live data. I've said it on the podcast: if you're big enough, you should have written a ChatGPT plugin *yesterday*. It's like the early days of mobile apps – those who got in early reaped outsized gains. There's a first-mover advantage right now in being the brand that *is* discoverable via conversational AI. Data plumbing plays a role because to build these plugins or feed these AI, you need clean, structured data (harking back to why a PIM is crucial – garbage in, garbage out for AI too). You may also need to set up new integrations – for example, using a framework like **LangChain** to connect your internal data sources to an LLM, allowing more sophisticated queries or internal assistant bots.

On the personalization side, **AI-driven personalization** can use the CDP data to do things like predictive recommendations, churn scoring, dynamic pricing optimization, etc. A simple win is using AI to decide the best content or offer for each customer in marketing emails – feeding the CDP profile into an AI model that chooses from a few variations. Some CDPs and marketing automation tools have this baked in; if not, you can integrate something custom. The key is that without unified data, your AI is flying blind or only seeing a piece of the picture. With unified data, even relatively straightforward machine learning models can yield powerful insights (like segmenting customers into clusters for targeting, or predicting who is likely to buy what next).

Let's not forget privacy and trust here: when collecting and unifying all this customer data, be transparent with users and allow them control. Modern consumers appreciate personalization but not at the expense of feeling spied on. A CDP should help manage consent (many do), and any AI usage of customer data should be done carefully and within ethical bounds.

Alright, we've covered a lot of ground on data plumbing components. To make it more concrete, let's walk through how different types of businesses might approach this journey, because a side-hustle B2B wholesaler is not the same as a booming direct-to-consumer brand.

INTEGRATION PLAYBOOKS FOR DIFFERENT BRAND ARCHETYPES
Every business has unique needs, but I often see e-commerce

companies fall into a few archetypes when it comes to data integration needs. Let's discuss three common ones – and some playbook tips for each – so you can see which sounds like you.

SIDE-CAR B2B BRAND (WHOLESALE WITH A NEW ONLINE CHANNEL)

Profile: You're a traditional B2B company (maybe wholesale, manufacturing or distribution) that's been doing business via phone, fax, sales reps, and maybe EDI, and you've recently launched an e-commerce site as a "side-car" to your existing operations. Typically, you already have an ERP that's the lifeblood of the business (managing POs, inventory, customer-specific pricing, etc.). The e-commerce site is often the newcomer and might not yet be fully trusted by the old guard internally.

Data Plumbing Focus: *Inventory and pricing integration* are usually the biggest deals here. B2B catalogs are often huge and complex, and pricing can be customer-specific (like different contract prices per client). The ERP is likely the master of those prices and available inventory. The last thing you want is a B2B customer logging in to the new site and seeing the wrong price or, worse, ordering something that isn't actually in stock – that could jeopardize long-standing contracts. So priority one is syncing those customer-specific catalogs and stock levels to the site. Initially, some Side-Car B2Bs launch e-com with a simple nightly sync – e.g., uploading a price file and inventory file once every 24 hours. But as customers start using it regularly, you'll need to move to at least hourly or real-time updates for inventory, and very frequent updates for pricing (especially if pricing changes often or if you allow quote requests online).

Integration Approach: I often recommend these businesses *start by integrating a few key things* rather than everything at once. For instance, **crawl** with just pulling inventory and pushing orders. Maybe at first, when an order comes in online, you simply email it to a sales rep or have someone manually enter it into the ERP (that's okay on day 1). Meanwhile, set up a daily product catalog sync from ERP to site so all SKUs and prices are up to date each morning. Then **walk**: automate the order push into ERP (so web orders appear in ERP within, say, 15 minutes) and increase inventory update frequency. Perhaps use an iPaaS that has a connector for your specific (often archaic) ERP – for example, a lot of B2B folks run things like Epicor, Sage, or Microsoft Dynamics; there are connector tools and middleware specifically for these. By the time you **run**, you ideally have *real-time pricing and inventory*. That might mean, for customer-specific pricing, the website calls an API to the ERP on the fly when the customer logs in (or caches it for their session). Or you pre-load the site daily with all price lists but then handle updates incrementally. For inventory, maybe you implement event-based sync: when an order is invoiced in ERP or stock is received, it triggers an update.

Pitfalls and Tips: Many B2B ERPs are older and not API-friendly (I've been in the trenches with those Epicor Eagles and AS/400-based systems). In such cases, consider a small **middleware service** that can speak the ERP's language (be it flat files, ODBC databases, etc.) and translate to modern API calls. This might be custom dev, but it's worth it to decouple the old system from your shiny new website. Also, B2B tends to have *more complex data* – like units of measure conversions, multi-warehouse availability, and lots of legacy data quirks – so invest time in data cleanup and mapping. One client of ours in B2B had multiple addresses and contact persons per customer in their ERP; when we integrated to the site, we had to decide how to handle that (we ended up creating sub-accounts on the website for each contact). These design decisions impact how smooth your integration will go. And don't forget training your internal team – if the CSRs or sales reps don't trust the e-com site's data, they'll tell customers to avoid it. Demonstrating that "the website always has up-to-date info" builds internal buy-in, which is crucial for adoption.

Quick case example: We worked with a pipe supply wholesaler (the one from the opening story) who initially had a very rough sync – one that failed, leading to the oversell. After that wake-up call, we moved them to a schedule of *15-minute inventory syncs* during business hours and immediate order posting to the ERP. Ultimately, we got fancy and did an on-demand check: just before finalizing checkout, the site would do a live call to the ERP to ensure stock was still available for those big B2B orders. That extra validation saved their hide a few times when multiple large orders came in on the same SKU. It's that kind of belt-and-suspenders approach that makes e-commerce trustworthy for B2B buyers who absolutely *need* what they order to show up as promised.

OMNICHANNEL RETAILER (BRICK & MORTAR + ONLINE)

Profile: You operate both physical stores and an e-commerce site – welcome to the omnichannel club! Your data plumbing has to connect in-store systems (which might include a retail POS, an ERP, maybe a separate inventory system or warehouse management) with your online platform. Customers expect a unified experience: buy online pickup in store (BOPIS), buy in store ship to home, consistent pricing and promotions, unified loyalty program, etc. This archetype often has legacy tech in stores (some retailers are running decades-old systems in their brick & mortar because ripping them out is like open-heart surgery). The e-commerce arm might be newer but has to live with those legacy constraints.

Data Plumbing Focus: The mantra here is **"single view of inventory, single view of customer."** You want a real-time global inventory if possible – knowing exactly how many units are in which store and which warehouse, and exposing the right piece of that to customers online. For example, if you do store pickup, your site needs to show per-store availability. That requires tight integration with the store inventory/POS systems. Likewise, if you allow ship-from-store to fulfill online orders, your e-com system needs to be in constant communication to decrement store stock when an online order is allocated to that store. On the customer side, if you have a loyalty or CRM in store, integrating that with the site (or feeding all of it into a CDP) is key so that a customer's purchases and returns are tracked across channels.

Integration Approach: A common pattern is implementing an **OMS (Order Management System)** that sits in the middle and orchestrates orders and inventory across channels. Systems like Salesforce Order Management, Oracle NetSuite, or even custom-built ones act as the traffic cop: when an order comes in online, the OMS decides "fulfill from e-com warehouse or from nearest store?" and updates inventories accordingly. If you go that route, your data plumbing shifts a bit – the OMS becomes the hub, and you integrate *it* with everything (ERP, e-com, POS). If you don't have a fancy OMS, you can still achieve a lot with point-to-point integrations: e.g., have the e-com platform pull store stock levels from the ERP or POS system's database periodically, and have in-store sales subtract from a common inventory pool that the website also sees.

For omnichannel, I advise aiming for **real-time inventory sync** between stores and online – it's just too risky otherwise. If a store sells the last unit of an item and your online inventory isn't updated until tonight, you could easily sell that "ghost" inventory online in the afternoon and then have to disappoint an online customer or frantically transfer stock. Many retailers solve this by **reserving a buffer** per store for online (like, don't sell the last 1 unit via BOPIS), but a better solution is just to have instant updates. Modern cloud-based POS systems (like Shopify POS, Square, etc.) do this naturally. Legacy ones… not so much, and that's where middleware can again come to the rescue.

Pitfalls and Tips: *Integration complexity.* That's the biggest challenge. You might have an ERP that handles purchasing and warehouses, a POS that handles store sales, a separate e-comm platform, maybe an old loyalty database – getting all that to sing in harmony is non-trivial. It often requires a multi-step project. Start with the most customer-facing pain point. In a lot of cases, that's **inventory visibility**. If BOPIS is big for you, invest early in integrating store inventory with the site (even if it's something crude like an API that the POS vendor provides or a script that dumps store stock to a file which the site imports every 5 minutes). I've done exactly that hack: a script grabbing inventory out of a POS database every few minutes and syncing to the website – not elegant, but it prevented oversells.

Another tip: **unified SKU system**. Ensure that the product identifiers (SKUs, UPCs) are the same in your store system and your online system. If they aren't, create a mapping table and keep it updated. I've seen cases where stores had a different SKU for an item than the website due to legacy reasons – that complicates everything from inventory sync to sales reporting. Clean that up as part of your plumbing initiative.

Also, consider the **customer data loop**. If someone buys in store, can they look up that order in their online account? That's a nice-to-have that requires feeding store sales into the e-com or account management system (often via the CDP or CRM). Even if you don't surface it to the customer directly, having all that data in one place means your marketing team can do omnichannel campaigns (e.g., segment customers who browse online but buy in store). So feeding offline data into your CDP or at least into your e-com customer profiles is valuable.

Case in point: We integrated an omnichannel farm and home retailer's systems so that their e-commerce site could show store-level inventory and enable pickup-in-store. Their ERP (Epicor Eagle, again) was used by all their physical stores to manage stock, but it wasn't built for real-time queries. We built a lightweight service that periodically pulled inventory from Eagle and updated Magento's stock numbers, and also flagged when inventory got below a threshold so we could disable pickup for that item at that location. Over time, we moved to an event-based approach – when a store sold an item, the POS would send an update to the central system that then pushed to the site. During holiday peaks, this meant the difference between 95% accurate inventory vs. 50% in the dark. And customers were happier – fewer "Sorry, we don't actually have that item you ordered for pickup" calls, which no one wants to make or receive.

Pure-Play Digital (Direct-to-Consumer eCom, No Physical Retail)

Profile: You're an online-only brand – maybe a digitally native vertical brand, a fast-growing Shopify or Magento store that started as a scrappy D2C idea and is now scaling up. You don't have the baggage of physical store systems, which is a plus, but you likely have a lean team juggling a lot of tools (eCom platform, 3PL for fulfillment, various SaaS apps for marketing, etc.). The challenge here is often **stitching together a unified back-end from many SaaS services** and knowing when to introduce systems like an ERP or PIM since nothing is dictated by legacy – it's all up to you.

Data Plumbing Focus: Early on, a pure-play might run everything from the e-commerce platform itself (it's your product catalog, inventory tracker, order system, and maybe even does some basic customer info). But as you grow, you'll hit limits. Common triggers are: your SKU count grows too big or your content needs too rich, so you need a

PIM; or your volume grows and you need an ERP or at least an order management system to handle procurement, accounting, etc.; or you start selling through multiple channels (your own site, plus Amazon, plus maybe a pop-up shop or wholesale) and suddenly multi-channel coordination becomes an issue. So for pure-plays, the question is **when and how to implement new components** of the stack and integrate them.

Integration Approach: One benefit of being pure-play digital is you're often using more modern, API-friendly tools from the start. Say you're on Shopify Plus with a 3PL for fulfillment – they likely have a solid integration already (via Shopify's built-in fulfillment APIs). If you add a SaaS PIM like Akeneo or Salsify, it will have connectors or apps for Shopify to sync products. So you can leverage those. The crawl stage might have been "just use Shopify for everything," then the walk stage is "introduce a PIM, maybe introduce Segment as a lightweight CDP for marketing data, use their turnkey integrations," and then run stage could be "bring on an ERP for financials/inventory, build a custom middleware to connect Shopify <-> ERP for real-time sync, and connect all the events to our CDP for fancy personalization."

For pure-plays, I often see **event-driven architectures** becoming appealing as they scale – using something like a message queue or event bus (e.g., Segment's tracking, or an AWS SNS/SQS pipeline) where every order, stock update, customer sign-up etc. emits an event that other services listen to. This can be a great scalable way to keep systems in sync without tight coupling. For instance, an order placed event gets picked up by your ERP integration process which then creates the order in ERP, another listener might send a Slack notification to your team, another might feed the data to your CDP. Modern commerce platforms and services often have webhooks precisely for this reason.

Pitfalls and Tips: The biggest pitfall for digital natives is sometimes *over-integrating or over-engineering too early*. Because you have so many shiny API-based tools available, there's a temptation to stitch together a Rube Goldberg machine of every best-of-breed service on earth by Month 3. I love composable architecture, but it has to be done with a plan (and the resources to manage it). My advice is to implement systems as needed, not just because. For example, if you're only on one channel (your site) and have 200 SKUs, you might not need a full PIM yet – a well-managed Google Sheet or the built-in catalog might suffice in crawl stage. But have a threshold in mind: e.g., "when we hit 500 SKUs or plan to add a second channel, we'll introduce a PIM." Same for ERP: maybe you can skate by with QuickBooks and a 3PL's inventory management for a while; but if you're struggling to track COGS, or you're doing pre-orders and need better inventory planning, that's when an ERP or more advanced inventory system should come in.

One tip: **choose tools that play nicely together**. Many SaaS e-commerce tools advertise their integrations – use that as a factor in picking software. If your email platform natively integrates with your e-com platform and your CDP, that's less custom work. Also, don't neglect the data warehouse aspect: a lot of pure-play brands now set up a cloud data warehouse (like Snowflake or BigQuery) and pipe all their data there for centralized analysis. That's more analytics plumbing than operational, but it's part of a mature data strategy and not too hard to do with tools like Fivetran or Stitch. It won't directly prevent oversells, but it will give you insights that might highlight process issues (like, "hey, we had 20 orders canceled last month due to stockouts – why?").

Example: One of our D2C clients scaled from $1M to $50M in revenue in a couple of years. In the early days, everything was on Shopify + some apps. As they hit about $10M, we helped implement a PIM because their product content was becoming a bottleneck (lots of new SKUs each season, translations for international sites, etc.). By $20M, we integrated an ERP (NetSuite in their case) to better handle purchase orders, multi-warehouse inventory across their US and EU fulfillment centers, and financial reporting. The integration we set up fed Shopify orders into NetSuite in near real-time and echoed inventory back to Shopify within minutes after a receipt in the ERP. We also set up Segment as a CDP around that time to unify their customer events (as they started doing some pop-up shops and expanding to marketplaces). Each step was triggered by a clear pain: content management pain -> PIM; ops and accounting pain -> ERP; marketing fragmentation -> CDP. The moral is, if you're a pure-play, keep an eye on where the pain is growing and address it with the right system at the right time – and *then* integrate it well.

No matter which archetype sounds most like your business, the endgame is the same: **a well-oiled data plumbing system that supports growth instead of hindering it.** And how do you know if your plumbing is well-oiled? You measure it.

Six Metrics to Monitor Your Data Plumbing Health

You can't improve what you don't measure, so I'm a big proponent of tracking metrics that indicate how healthy your integrations and data flows are. Here are six dashboard metrics I recommend monitoring:

Inventory Sync Lag: This measures the average (or max) delay between an inventory change in the source (e.g., warehouse / ERP) and that change reflecting on the website. For example, if you receive new stock at 10:00 AM and your site shows it in stock at 10:30 AM, that's a 30-minute lag. You want this as low as possible for critical products. Set a target (say, < 5 minutes for top SKUs, < 15 for all SKUs) and track it. If you see spikes, investigate what caused the delay (network issue? API down?).

Oversell / Stockout Rate: How often are you selling something you can't fulfill because of bad data? This could be measured as "percent of orders that later got backordered or canceled due to no stock" or a simple count of oversell incidents. It should be essentially zero in a healthy system. If it's not, tie it back to which integration or process failed and fix that. Oversells are the canary in the coal mine for inventory sync issues.

Product Data Completeness: A quality metric – what percentage of your active products have complete information (all required attributes filled, at least one image, etc.)? Your PIM or even your e-com platform might provide this. For instance, maybe only 80% of products have dimension weight info filled in – that's a gap that could lead to shipping cost errors. Track this and aim for 100% completeness on all critical attributes. If you see it dip when lots of new SKUs are added, it might mean your process for adding products is rushing things live without full data – an opportunity to adjust workflow.

New SKU Time-to-Live: The average time it takes from "product defined in ERP/PIM" to "product live on the website." If it currently takes 10 days and a lot of manual work, measure that and then try to bring it down with better integration and process. Maybe after a PIM implementation and some automation, you get it to 3 days. This metric ties together cross-functional effort (it's part process, part integration), but it's a great one for demonstrating the value of your data plumbing improvements to the business – faster time to market is something everyone appreciates.

Integration Error Rate: If you have any logging on your sync processes (and you should!), track the error rate. For example, number of order sync failures per 1000 orders, or number of product records that failed to import last week. Ideally, this trends toward zero as you iron out bugs. A spike in errors means something's wrong (e.g., a change in a data field that wasn't accounted for, or a service downtime). Monitoring error trends helps you catch issues proactively. Even if they're auto-retrying, errors can indicate latency or data issues that need attention.

Unified Customer Profile Coverage: If you've got a CDP or even just are merging data in some way, measure what percentage of customer records or interactions are unified. For example, if 70% of online orders can be linked to an existing profile that also contains in-store purchases, that's your unified coverage. You want this high, because it means your marketing and service teams have the full context. Low coverage might mean data from one channel isn't being integrated or identity matching (like matching emails to emails, or emails to phone numbers) isn't working well. Improving this might involve tweaking how you collect emails in-store, or how you merge guest checkouts with known accounts via machine learning or deterministic matching.

Depending on your business, there may be other vital signs – for instance, **API response time** for critical integrations could be one if you're very API-driven (slow API = slow updates). Or **manual touch count**: how many times per week does someone have to manually fix or intervene in a data process (you want that trending down to zero). But the six above cover a broad range of plumbing health indicators. Consider creating a simple "data ops" dashboard for your team that surfaces these. When something goes out of whack, you'll catch it early.

90-DAY ACTION PLAN: FROM AUDIT TO REAL-TIME NIRVANA

Alright, you might be thinking, "This is great info, but where do I actually *start*?" So let's lay out a high-level 90-day game plan to level up your data plumbing. Three months is a realistic timeframe to get a lot of foundational work done without it dragging on forever. Here's how I'd break it down:

Days 1–30: Audit and Map the Current State – Start by taking inventory (no pun intended) of all your current data flows and pain points. Map out each system and how data moves between them today. Document frequencies, formats, and who is responsible. For example: "ERP exports CSV of inventory daily at 2 AM, FTP to server, eCom platform imports at 3 AM." Do an end-to-end audit: products, inventory, orders, customer data, etc. At the same time, gather the horror stories: talk to customer service about recent oversells, ask the web team how often they have to manually adjust something, ask the ops team if there are any integration issues that keep them up at night. **Gather baseline metrics** where possible (from the six above or others) – how fast are things now, how many errors happen, etc. This first 30 days is also when you should question the status quo: Is every system here necessary? (E.g., do we actually use that old PIM or can we drop it? Or conversely, are we trying to make the ERP do something it shouldn't, like manage rich product content, and thus suffering?) If you have the capability, also use this month to implement some basic monitoring on key processes, so you have data to back up anecdotal issues.

Days 31–60: Quick Wins & Integration Roadmap – In the second month, tackle low-hanging fruit and plan the bigger moves. Quick wins might include: increasing a sync frequency (if something is daily, can you make it hourly without writing new code? maybe the tool supports it), fixing a data mapping (oh, that unit mismatch – let's correct the format now to stop errors), or setting up an alert for a known failure point. These small changes can already reduce risk (for example, switching that inventory update from daily to hourly might immediately cut oversells by 80%). Concurrently, develop your **longer-**

term integration plan. Based on the audit, decide what "run stage" looks like for you and what's needed to get there. This could be: selecting a PIM system and planning its implementation, deciding to implement a CDP (or simpler: deciding to pipe data into a data warehouse first as a step), identifying that an iPaaS solution could replace custom scripts and researching which one. It could also be the time to decide on any **new system acquisitions** – e.g., "we will adopt Akeneo PIM, timeline X" or "we need an OMS, let's evaluate options." By day 60 you should have a clear picture of what you're going to implement or improve, and have kicked off any necessary purchasing/procurement or gotten approvals for those projects.

Also in this period, start designing the **integration architecture** for the target state. It might be diagrams on a whiteboard or in Lucidchart mapping ERP → PIM → eCom → CDP flows, and identifying where middleware or connectors are needed. If you have development resources, they can begin prototyping one of the new integrations or setting up a staging environment for the new PIM or CDP to play with sample data.

Days 61–90: Build, Test, and Launch Iterative Improvements – The last 30 days is about execution. Implement the changes you planned. This could mean actually standing up that PIM and migrating product data into it, then connecting it to the website. Or writing the custom API integration between the website and ERP for inventory. Focus on one or two core integrations first – for instance, get real-time inventory syncing working end-to-end. Test it thoroughly: simulate orders, see if stock updates flow properly within the expected time. Use this time to also set up the **observability/monitoring** around the new flows. If you launch a new integration without monitoring, you're asking for trouble. Even a simple log file that says "synced 1000 products at 12:00, 0 errors" is fine, as long as someone or something checks it.

By around day 75 or so, start **training the team and flipping the switch** on new processes. If customer service or the warehouse team needs to use a new interface or expect changes (like "the eCom site will now show inventory by store – here's how to answer customer questions about it"), get them ready. Then, go live with the changes. Maybe you cutover from the old CSV import to the new API integration for orders – do it during a slow period and have all hands on deck to watch for issues.

By day 90, you should be seeing the benefits: fewer manual tasks, fresher data on the site, fewer customer complaints of wrong info. Take new measurements on those metrics you established. Hopefully, you can report something like "we reduced oversell incidents by 90%" or "product launch time went from 2 weeks to 3 days." Those are big wins. Don't forget to communicate these wins to leadership – improving data plumbing might be behind-the-scenes work, but it directly impacts revenue and customer experience, so it's worth tooting your horn a bit.

One more thing in the 90-day plan: lay out the **ongoing mainte-nance routine**. Assign owners for each integration (who gets the alert if it fails?), schedule periodic reviews (maybe monthly check of logs for silent issues, quarterly review of whether we need to tweak something as business needs change). And keep an eye on the horizon – maybe AI integration isn't urgent in this 90 days, but you know it's coming, so keep those future-facing items (like building a ChatGPT plugin or introducing more AI in workflows) on your backlog and revisit them once the basics are solid.

Congratulations – if you follow this plan, in three months you'll have gone from rusty pipes to a modern flow. It's the difference between bailing water with a bucket versus having your plumbing automatically pump it to where it needs to go. Your team will thank you (less firefighting!), and your customers will reward you (with their dollars and loyalty).

Data plumbing might not be the sexiest topic at first glance, but hopefully I've shown that it's absolutely pivotal to sustainable e-commerce growth. The brands that nail this – that can trust their data, integrate new tech quickly, and leverage unified information – are pulling ahead of those that are constantly reacting to the latest leak or error. And the best part is, once your data plumbing is in place, you can build amazing things on top of it: personalized AI-driven shop-ping experiences, lightning-fast fulfillment, you name it.

Speaking of lightning-fast – in the next chapter, we're shifting gears from back-end plumbing to front-end performance. Now that your data is flowing smoothly behind the scenes, it's time to make sure your **site is equally smooth and fast for customers**. We'll dive into Chapter 3.4: *Site Speed & Core Web Vitals*, where we tackle how to get your pages loading in a snap and why Google cares so much about those performance metrics. Speed is the next battleground for e-commerce conversion, so you won't want to miss it. See you there!

3.4: SITE SPEED & CORE WEB VITALS: THE REVENUE MULTIPLIER

Boom. Your brand just got a shout-out from a TikTok influencer with 5 million followers. Traffic spikes, hundreds of eager shoppers click the link… and then they wait. And wait. Six seconds of blank white screen. In internet time, that's an eternity. The influencer refreshes on her livestream, rolls her eyes, and says, *"Looks like they weren't ready for us, guys."* The moment fizzles, and so do your sales. This isn't a far-fetched horror story—it's everyday life for sites that ignore speed. In fact, a one-second delay in load time can drop your conversion rate by about **9%**. Think about that: Speed isn't some geeky tech metric; it's a direct lever on your revenue and profit margin. Every extra second is literally money leaving your pocket. **Site speed is a margin killer or multiplier, depending on how seriously you take it**.

Before we dive into tactics, let's level-set on why this still matters **today**. We've been hearing about site speed for over a decade, but many ecommerce operators still treat it as an afterthought. Here's the reality: shoppers are more impatient than ever, and Google's algorithms haven't gotten any more forgiving either. About 70% of the *perceived* speed of your site comes from the *user experience* layer — things like how quickly something appears on screen and whether the page *feels* snappy. In other words, front-end performance and responsiveness shape the majority of a user's speed impression. And that impression doesn't just influence bounce rates; it bleeds into your SEO rankings, your conversion rates, and even your brand reputation (as our influencer example shows). Site speed isn't a "tech team problem." It's a **business fundamentals** problem.

So in this chapter, we're going to get very practical about web performance for ecommerce. No fluff, no endless theory—just real-world strategies to make your site faster and your customers happier (all in plain English). I'll share a crash course on Core Web Vitals (Google's user-centric performance metrics), a five-step "leak test" you can run on your site to find where you're losing speed (and money), and a toolstack cheat-sheet from free to enterprise solutions. We'll look at quick case studies of speed wins, highlight common pitfalls (so you don't shoot yourself in the foot), and even lay out a one-week action plan to start boosting your site speed **right now**. By the end, you'll see why optimizing site speed is one of the highest ROI moves you can make in ecommerce—and how to embed a culture of speed in your team. Let's turn that performance from a leaky bucket into a revenue firehose.

WHY SPEED AND CORE WEB VITALS MATTER

If you've been in the SEO or dev game for a while, you might remember when Google rolled out the **Core Web Vitals** as ranking signals back in 2021. A few folks said, "Meh, it's just a tiebreaker." Others treated it like the holy grail. What's the truth? Do these performance metrics *really* impact your bottom line? The short answer: **Yes**—but not always in the ways you'd expect.

First, let's talk SEO. Google *does* use Core Web Vitals (CWV) as part of its ranking algorithm for search results. It's not the *top* factor (relevance and content quality still rule), but think of CWV as a tiebreaker. If two pages have equally relevant content, the faster, better-performing page is likely to rank higher. In fact, Google has stated outright that Core Web Vitals act as a tie-breaker between similar content, and crucially, you only get an SEO boost if you're passing **all**three CWV metrics – there's no partial credit for doing well on one and failing another. It's basically pass/fail. If your site nails two metrics but has, say, a poor LCP, Google considers that a failed page experience overall. In competitive markets, that can be the difference between showing up on page 1 or page 2.

Secondly, Google's page experience measurements are **real-user metrics**. This is super important: the scores that count for rankings come from actual users loading your pages (via Chrome's anonymized data), not from a synthetic lab test. That means no matter how many times you run Lighthouse or PageSpeed Insights in dev, those *lab scores* won't directly influence your SEO. Only the **field data** (what Chrome users experience at the 75th percentile) matters. So if you've ever wondered why your site's "Performance: 90" in Lighthouse hasn't budged your SEO, that's why. Lab tools are for diagnostics; real users are the truth. It also means optimizing for **consistency and broad user conditions** is key – you want virtually everyone, on all kinds of devices and networks, to get a fast experience, not just a pristine lab scenario.

Mobile performance remains the priority. Google shifted to mobile-first indexing years ago and continues to weigh mobile page experience heavily for rankings. If your mobile site is slow, your SEO will suffer on mobile searches – and since most searches are mobile, that's your primary battlefield. I've seen clients scratching their heads as to why their mobile traffic lags their desktop, only to realize Google down-ranks them on smartphones because their mobile load times stink. *Don't let that be you.* Make sure to test on mobile devices and even throttled networks.

Beyond the pure SEO angle, here's something even more straightforward: **speed converts**. Faster sites deliver a better user experience, which leads to happier users, which leads to more dollars in the till. We already dropped the stat that a 1-second improvement can bump conversions ~9%. But I'll back it up with a real-world example: Renault (the car manufacturer) reported that cutting their Largest Contentful Paint by just 1 second reduced their bounce rate by 14% and *increased conversions by 13%*. Think about that – 13% more sales without spending a penny on more ads or traffic, just by speeding up the site. That's the "revenue multiplier" effect of site speed. I could rattle off more: Vodafone improved LCP by 31% and saw 8% more sales, Swap-pie's performance optimizations drove 42% higher mobile revenue, and so on. The patterns are consistent.

The key point: Core Web Vitals (and site speed generally) remain critical because they sit at the intersection of **SEO, UX, and conversion**. Google rewards good CWV with better visibility (especially when all else is equal) , and users reward good CWV with deeper engagement and purchases. It's a double win. Conversely, if you ignore performance, you might get hit twice: slightly lower rankings *and* a leaky conversion funnel for the traffic you do get. That's why I call site speed the "margin multiplier" – it can either quietly erode your margins or boost them significantly.

One more angle: **real user monitoring**. Modern ecommerce teams are increasingly instrumenting their sites with RUM tools (Real User Monitoring) to track performance continuously. If you haven't looked into this, it's time. Google Analytics 4, for example, now surfaces some basic Core Web Vitals data for your site via the CrUX dataset, and services like RUM Vision specialize in giving you a live pulse of your site's speed as experienced by actual visitors. This means you can move beyond occasional testing and catch issues in the wild – like if a new deploy slowed down your homepage for real users, or if users on a certain device type are struggling. In 2025, savvy teams treat performance metrics just like uptime or error rates: something to monitor in real-time. We'll talk more about tools for that, but the big picture is that speed has graduated from a one-off project to an ongoing operational priority.

Bottom line: If someone on your team asks, "Do we really need to worry about Core Web Vitals and speed? Isn't our site 'fast enough'?" — you have my full permission to show them the door. (Or, more gently, show them the data above.) Speed is *absolutely* still a competitive advantage and a revenue factor. And unlike many marketing levers, it's under **your** control. Let's use that control.

CORE WEB VITALS CRASH COURSE (AND THE AMAZON EXAMPLE)

Time to get hands-on. Google's **Core Web Vitals (CWV)** are the three amigos of site speed metrics, each measuring a different aspect of user experience: **loading, interactivity**, and **visual stability**. They sound technical, but I promise they're straightforward once you get the concept. Let's break them down in plain language. I'll also share a quick reality check story about Amazon that illustrates why chasing a perfect "score" isn't the goal – **real user experience is**.

First, here's a cheat-sheet of the Core Web Vitals, their "good" thresholds, why they matter for your users, and a couple of quick fixes for each:

Largest Contentful Paint (LCP)

Target: < 2.5 seconds

Time to show the main content (e.g. hero image). Slow LCP means users stare at a blank or half-loaded page – they bounce out.

Optimize your hero image (compress & resize); implement lazy-loading for images **below the fold** (so offscreen images don't delay the first screen).

Interaction to Next Paint (INP) *(replaces FID)*

Target: < 200 milliseconds

How quickly your site responds when a user tries to interact. If people click and nothing happens for >0.2s, the site feels laggy.

Prune heavy JavaScript; defer non-critical JS so it loads later; use browser APIs to do work off the main thread. Essentially, trim the JS bloat that stalls interactivity.

Cumulative Layout Shift (CLS)

Target: < 0.10

Measures how much the page jumps around during load. High CLS = elements moving unexpectedly (user goes to click a button and it moves – infuriating!). This hurts trust and usability.

Reserve space for elements (e.g. set image dimensions or CSS aspect ratios); avoid sudden pop-ups or banners that push content; load ads in iframes with fixed size. Keep the layout stable.

Those are the basics. Hit all three of those thresholds (often referred to as "green" scores), and your site is officially providing a good user experience from a performance standpoint. It also means you're passing Google's Page Experience test in Search Console, which ties back to that SEO boost we discussed.

Now, a quick note on the newcomer: **INP (Interaction to Next Paint)**. This metric is replacing the old FID (First Input Delay), because it gives a more complete picture of how responsive your page is over its whole lifecycle. FID only measured the very first interaction delay, whereas INP observes many interactions and picks a sort of worst-case (or 98th percentile) latency. In practical terms, just know that Google wants sites to respond *nearly instantly* to user inputs – 200ms or less is the goal. If your site has a lot of long tasks (usually from JavaScript) hogging the main thread, your INP will suffer. We'll fix that by cutting script bloat in a minute.

To make this less abstract, let's use a real example: **Amazon.com's mobile site**. Everyone knows Amazon; they have the budget and talent to optimize like crazy. You'd think they'd score 100/100 on Google's performance tests, right? Wrong. Amazon is a poster child for the idea that *lab scores aren't everything*. When you run Amazon's homepage through Google's Lighthouse (the engine behind PageSpeed Insights), it usually scores in the red – around a 46 out of 100. Ouch. But hold on: if you look at Amazon's actual Core Web Vitals, they're pretty darn good. In one analysis, Amazon's Largest Contentful Paint on mobile was about **1.4 seconds**, their First Input Delay (FID) – roughly analogous to INP – was just **24ms**, and their CLS was a tiny **0.01**. In other words, real users are getting a fast, responsive, and stable experience. So why the heck is their Lighthouse score 46 (which normally indicates poor performance)?

It comes down to how Lighthouse scoring works: it penalizes things like Total Blocking Time (TBT – heavy scripting) and uses a simulated device/network. Amazon's site is loaded with JavaScript (for third-party trackers, internal tools, etc.) that in a synthetic test environment trip the alarms. In one test, Amazon's Total Blocking Time was over 1.5 **seconds** (1500ms), and First Contentful Paint (when something *first* appears) took 3.6s in the lab. Those are not "good" numbers, hence the lousy 46 score. But Amazon has engineered the site so that users still *perceive* it as fast: the important content (LCP) shows up quickly (1.4s is quite good), nothing jumps around (CLS 0.01), and if you tap something, it reacts instantly (sub-50ms) in most cases. The heavy scripts are loading in the background and not blocking the critical path as much as the raw metrics imply. Also, Amazon's real users likely have certain resources cached, etc. In practice, Amazon ranks fine and sells a ton, proving that **passing CWV is more important than** acing **Lighthouse**. As one RUM tool expert put it: "Lighthouse scores don't really matter at all; it's really the Core Web Vitals that count".

The takeaway for you: **Focus on the Core Web Vitals and real-user experience, not chasing a perfect Lighthouse score.** If you can get that elusive 100/100, great, but don't do it at the expense of real UX. I've seen teams cheat the Lighthouse score (more on that in a bit) or optimize weird things just to please the test, with zero benefit to actual users. Instead, use Lighthouse and PageSpeed Insights as *tools to identify issues*, but measure your success by CWV and, ultimately, by improvements in conversion or engagement metrics.

Alright, now that we know what LCP, INP, and CLS are (and that we care about them in the real world), it's time to actually improve them. Let's move to a step-by-step audit that I like to call **the Five-Step Speed Leak Test**.

THE FIVE-STEP SITE SPEED "LEAK TEST"

Think of your website's performance like a bucket of water. Every leak – every unnecessary script, image, or slow server response – is water (revenue) seeping out. Over time, small leaks turn into big losses. I developed a five-point "leak test" that helps systematically find and fix the most common speed leaks on an ecommerce site. We actually walked through these tests on an episode of my podcast, and the response was great because it turns an overwhelming topic into a simple checklist. Even if you're not a developer, you can run many of these tests yourself or in tandem with your dev team. Let's go through them:

Test 1: Speed Baseline – Measure, Measure, Measure. You can't improve what you don't measure. The first step is to benchmark where you are. Run your homepage and a key product page through **Google PageSpeed Insights** (which uses Lighthouse) or **Lighthouse in Chrome DevTools**. Look at the Core Web Vitals results and note which metrics are failing or barely passing. Is your LCP 3.8s on mobile (fail)? Is your CLS 0.25 (fail)? PageSpeed will tell you. Also note the *diagnostics* it gives – PSI is great at listing the biggest contributing factors (e.g. "Image XYZ is 1.2MB, compressing could save…", or "Eliminate render-blocking resources"). Save these reports. The goal here is to get a baseline "speed scorecard" for your site. **If any Core Web Vitals are**

in the red, that's a glaring leak to fix. Even if they're yellow (needs improvement), mark it down. This baseline will not only guide your action plan but also give you something to compare against after you make improvements (nothing more satisfying than seeing those red metrics turn green). In our experience, many sites find at least one CWV metric failing – and that's a prime indicator that a front-end overhaul like Hyvä (for Magento users) might be a game-changer. But even before considering drastic changes, you can likely fix a lot of issues manually, which brings us to the next tests.

Test 2: Dev Velocity – How fast can you fix fast? This one is a bit meta: it's not about your site's speed, but your **team's speed** in improving the site. Why does that matter for performance? Because if every tiny change takes a dev team 3 weeks and a code deployment, you're less likely to tackle performance issues in an agile way. I've found that sites on modern, modular front-ends (like the Hyvä theme for Magento) allow developers to work much faster – often cutting front-end development time by 30% or more compared to older frameworks. That means if a marketing team needs to swap a banner or add a new content section, it's a trivial change, not a "open a Jira ticket and wait" ordeal. To assess this, look at your recent development history: how long does it take to go from "ticket created" to "change live in production" for front-end updates? If simple things (like updating a homepage banner) require deep developer involvement or have long lead times, your front-end stack might be a performance bottleneck in itself. We once audited a merchant who had to **pay a developer** to swap out the homepage hero image and then wait a full sprint for it to go live – that's a sign your theme/platform is overly cumbersome. Ideally, your team should be able to iterate quickly on front-end improvements. If not, consider investing in a lighter-weight front-end framework. (Shameless plug: that's one reason we love Hyvä for Magento – in our projects at Creatuity, we've seen front-end build times and costs drop substantially, sometimes **40-50% lower in year-one costs** versus a heavier solution.) Even outside of Hyvä, any architecture that streamlines development (like not needing to rebuild an entire JS bundle for a small change) will pay dividends. In short, speed of site and speed of development go hand in hand. Measure your dev velocity, and if it's slow, mark that as a "leak" to plug with better tools or processes.

Test 3: Script Bloat Audit – Hunt down third-party villains. If I could only do one thing to improve an ecommerce site's speed, it might be this: audit all the JavaScript and third-party tags loading on the site. I guarantee you'll find junk to remove. Over the years, marketing and analytics scripts accumulate like plaque in an artery – and they can choke your site's performance. Tools like **BuiltWith** or the network waterfall in Chrome DevTools will list all the external scripts loading. Make an inventory and ask, *"Do we really need this? What value does it provide relative to its cost in speed?"* I've seen sites where removing one unused script shaved **3 to 7 seconds** off load time and instantly bumped conversion by 5%. For example, one client's previous agency had installed an error-tracking script that pinged a service every time an error occurred – sounds smart, except the agency had long since canceled their account with that service. The script was effectively doing nothing except adding ~2–3 seconds to the page load for every user ! We toggled it off in the tag manager, and boom: faster site, no negative impact on functionality. Sometimes it really is that easy – a single admin panel switch to turn off a feature can remove a huge drag on performance. When you run PageSpeed Insights, check the "Diagnostics" section – it will often flag the worst offenders, like "Third-party script XYZ took 4s". In my experience, **over half** of the sites out there have at least one third-party script that either isn't used anymore or isn't worth its weight. Do a culling. Also, consider lazy-loading or deferring scripts that aren't needed immediately. Many chat widgets, for instance, can be loaded after the main content. By pruning and optimizing scripts, you'll not only improve metrics like INP (since the browser isn't busy executing as much JS) but also reduce Total Blocking Time and often improve LCP (less competition for resources). To put it simply: every script you remove is one less thing slowing your site down – and likely one less thing breaking, too. Take a hard look at each one. This step is free and can be done in a day or two, and it's not uncommon to see *huge* gains. (Quick tip: if you're non-technical, get your dev to generate a **coverage report** or performance audit in Chrome – it visually shows how much unused code is in each script. The results can be pretty shocking.)

Test 4: Media & Asset Diet – Cut the fat from images and videos. After JavaScript, the next biggest bloat factor is usually media files – images, videos, fonts, etc. So the fourth test is to audit your site's "diet." Are you serving over-sized images? Are you loading a 4K product photo where a 800px image would suffice? Are you using PNGs where JPEGs or WebPs would be more efficient? This is low-hanging fruit. Tools like PageSpeed will directly call out "Efficiently encode images – potential savings: X MB" if you have issues. I suggest identifying your top 5–10 heaviest pages (home, top category, etc.) and listing the media files they load. Then optimize **each one**: compress them, resize to the actual display dimensions, and consider next-gen formats (WebP/AVIF). Modern ecom platforms often have image optimization plugins or it can be done as part of your build process. Also, make sure you've got proper **caching** headers on images and assets so returning visitors load them from cache. It's painful how often I see an otherwise modern site that just forgot to enable production mode or a CDN, resulting in zero browser caching – meaning repeat visitors are re-downloading the same large images every time. Don't let that happen. Enable full-page caching or at least static asset caching at the server level (most platforms like Magento, Shopify Plus, etc. have this – sometimes it's a simple config flag). If you're self-hosting, verify that HTTP caching headers (Cache-Control) are being sent for static resources. On the media diet, also consider your *video* strategy: loading a big background video? Make sure it's compressed and perhaps doesn't autoplay on mobile (users on cellular will thank you). Lastly, leverage a **Content Delivery Network (CDN)** for your static assets. A CDN will serve your images, JS, and CSS from servers nearest to your user, reducing latency. It's usually a turnkey add-on (Cloudflare, Akamai, Fastly, etc.) and can instantly shave 50–200ms off load times for distant users, sometimes more. It also offloads traffic from your origin server, which can improve stability. In short, make your site lean: only the necessary bytes, in optimized form, delivered from as close to the user as possible. This "diet" phase often improves **LCP**significantly because images are a common LCP element – a smaller image = faster LCP. And it can help CLS too (ensuring images have dimensions set prevents jank). Optimize images and assets, and you'll see the speed gains immediately.

Test 5: Real-Time Monitoring – Set up eyes on glass. The last test is more of a habit to develop: implement ongoing performance monitoring. This isn't a one-and-done fix; it's equipping your team to catch regressions and understand performance continuously. There are a couple ways to do this. On the *free* end, as I mentioned, Google Analytics 4 now integrates with Core Web Vitals data (if you link your Search Console) and there are community tools like **PerfBudget.js** that you can add to your site to log performance data. These can alert you if, say, your LCP suddenly spikes above 4s for real users. On the paid side, tools like **RUM Vision** (for real-user monitoring) are affordable and can give you granular insights – e.g. you can see your P75 LCP for users in different regions or browsers. For a more holistic approach (especially if you have a complex site/app), consider an APM tool like **New Relic or Datadog** which offers both real user monitoring and synthetic monitoring. New Relic, for example, can track your site's Apdex (user satisfaction) and catch Javascript errors or slow API calls in real time. I've worked with clients using New Relic's browser monitoring to great effect – you can literally watch a live feed of user experiences and see, for instance, "Oh, 5% of our users on Chrome had an INP above 500ms yesterday – what happened?" Maybe a deploy introduced a blocking script; with monitoring, you'll catch it and fix it before it hurts too much. The point of this step is to **treat performance as an ongoing KPI**, not a one-off project. Set thresholds (like "LCP should stay under 2.5s, CLS under 0.1") and monitor them. Many tools let you set up alerts, so if something goes out of bounds, you get an email or Slack ping. For instance, I recommend setting up a weekly or daily Core Web Vitals report via Google Looker Studio (formerly Data Studio) pulling from CrUX or GA4. That way, everyone has visibility. Real-time monitoring also extends to uptime and load testing – consider running quarterly load tests to ensure your infrastructure can handle peak traffic without slowing to a crawl (nothing like learning about a database bottleneck during Black Friday). By establishing monitoring, you create a feedback loop: you'll know if your speed improvements are truly working and catch new "leaks" as they appear.

Run these five tests on your own site. By the end, you should have a clear picture of where your speed leaks are: whether it's failing CWVs, slow dev cycles, bloated scripts, heavy images, or lack of monitoring. Now, let's equip you with some tools to actually plug those leaks efficiently.

Tool Stack & Cost Cheat-Sheet for Site Speed

Improving site speed can feel overwhelming – there's a *ton* of advice and tools out there. To cut through the noise, I like to break the tool landscape into a few categories based on needs, and map them to budget levels. Whether you're running on a shoestring or have enterprise backing, there are solutions for you. Here's a quick tool stack cheat-sheet:

Synthetic Testing *(Lab monitoring)*

Lighthouse (Chrome DevTools), PageSpeed Insights; WebPageTest.org (detailed waterfall analysis)

SpeedCurve or Calibre ($$) – nicer dashboards, continuous monitoring

Catchpoint or Dynatrace ($$$) – enterprise-grade synthetic testing globally

Real User Monitoring (RUM) *(Field data)*

GA4 + CrUX (Core Web Vitals in Google tools); **PerfBudget.js** (open-source script to log CWV)

RUM Vision or SpeedCurve's RUM ($) – affordable RUM focused on CWV and UX metrics

New Relic Browser, Datadog RUM ($$) – deep analytics, slice by user segments, with alerting and integration into full APM suites

Front-End Overhaul *(Framework/theme)*

Hyvä theme for Magento (one-time license, highly cost-effective for Magento sites to get green CWVs); or lighter Shopify themes, etc.

Shopify **Hydrogen** (for headless storefront on Shopify, developer-friendly but requires custom build)

Fully custom React/Vue SPA (Next.js, etc.) – maximum flexibility but highest cost/complexity (often only justified for very large teams)

A few notes on the above: If you're just starting, stick to the free tools first – Lighthouse and WebPageTest are incredibly powerful for diagnostics. WebPageTest even lets you simulate different devices and speeds and gives filmstrips of your load. For RUM, definitely use the free data you have (Search Console's Core Web Vitals report, GA4's PageSpeed insights). You can always step up to paid RUM if you need more granular data.

On the front-end overhaul: I mention Hyvä (pronounced "hoo-vuh") for Magento because it's been a **game changer** for sites on the Magento/Adobe Commerce platform. It replaces the bloat of the old Luma theme with a streamlined, modern stack (Alpine.js, Tailwind CSS) and routinely delivers perfect or near-perfect CWV scores out of the box. We've seen Magento merchants go from failing CWVs to all-green literally overnight by implementing Hyvä, with much less development cost than a full headless rebuild. If you're on Magento and struggling with speed, Hyvä is probably option #1 to evaluate. For Shopify folks, Shopify's Hydrogen is a newer approach to go headless using React – it can yield great performance if done right, but it's not a magic bullet (and it requires React developers). Many mid-size brands do well with just optimizing their Liquid theme or using something like Shopify's Online Store 2.0 features combined with good old image/script optimization.

Enterprise SPAs (single-page apps) built with custom frameworks can be blazing fast *if* you have the engineering prowess to optimize every ounce. But I'll be frank: I've seen many go the SPA route to "improve speed" only to end up with a slow, complex site because they underestimated the challenge. If you can achieve sub-2s loads with a simpler approach, do that first. Simpler is usually faster (both in performance *and* in how quickly your team can ship updates).

The table above should help you budget your performance initiatives. If you have zero budget, you can still accomplish a ton with free tools and elbow grease (we basically did that in the five-step test). If you have some budget, a service like SpeedCurve or RUM Vision can automate a lot of monitoring and make your life easier – think of them as giving you a performance dashboard to share with your team and track progress. And if you have enterprise resources, tools like Catchpoint or New Relic will integrate performance into your broader ops toolkit (though you might need a specialist to set up all the dashboards and alerts to get the most out of them).

One more honorable mention: **New Relic** and similar APMs often have both RUM and synthetics. In Episode 69 of the Commerce Today podcast, we talked with a developer about using New Relic to keep a site in good condition, catching errors and performance issues quickly. One benefit of such tools is they aggregate all your errors and slow transactions in one place, saving developers tons of time diagnosing issues. For performance, New Relic's Browser module can show you page load breakdowns by browser, geography, etc., and tie in business metrics. If you're at a scale where every 0.1s of latency might mean millions of dollars, these tools are worth every penny.

Now that you have tools in your arsenal, let's get inspired by a few quick **case studies** of speed improvements leading to real gains.

FAST WINS: THREE SPEED CASE STUDIES

Sometimes the best way to drive home the value of site speed is through real stories. I want to share three mini-case studies (anonymized, but based on actual scenarios I've encountered) that show the kind of results you can get from performance work. These are the "revenue multiplier" moments that make you realize all this effort is worthwhile:

The $5M Retailer and the Mystery Script – A mid-sized online retailer was perplexed by slow page loads (6-8 seconds on first load) and underwhelming conversions on a recent campaign. A quick audit revealed an old third-party script in the site's header, originally intended for A/B testing, that was still running on every page. The kicker: that A/B testing service had been deactivated months ago. So this script was literally doing nothing except eating up load time. We toggled it off in the tag manager (no code deploy needed) and immediately cut **3-4 seconds** off the page load for first-time visitors. The very next week, the retailer saw conversion rate climb roughly 5% (their analytics showed a jump from ~2.0% to ~2.1% overall). It makes sense – faster pages mean less drop-off during landing and product browsing. A 5% lift in CVR for a business doing $5M annually is substantial money. This was essentially free revenue gained, all by removing a single line of JS. The lesson: you might have one of these "mystery scripts" lurking; find it and kill it!

From Sluggish to Snappy – INP Improvement – An ecommerce brand in the fashion space had decent overall load times, but their interactivity was lacking, especially on mobile. Users would tap menu buttons or filters and the site would lag before responding. We measured their Interaction to Next Paint (INP) on mobile and found it around **300ms** at the 75th percentile – meaning a lot of users experienced a one-third second delay on interactions. That doesn't sound huge, but it is noticeable (and it violates the "under 200ms" guideline). Our team performed a **full script audit and optimization**: we identified heavy JavaScript functions running on page load (including an overly aggressive image carousel script and some live chat widget code that loaded before it was needed). We deferred what we could and trimmed what we couldn't (replaced a carousel with a lighter one, etc.). The result? The site's INP (p75) dropped to about **80ms** on mobile – basically instantaneous. And their mobile conversion rate, which had been consistently around 1.4%, climbed to ~1.45% (+4%) in the following month, holding steady as the new baseline. Now, many factors influence conversion, but the only major change in that period was the performance fixes, and user engagement metrics (like bounce rate and pages per session) improved too. The takeaway: improving responsiveness can directly impact users' propensity to continue and complete a purchase, especially on mobile where people have less patience for janky behavior. If your site feels sluggish in reacting to input, you *will* lose some fraction of customers who get frustrated.

Hyvä Rollout – 14 Days to Green and 11% More Revenue – This is a story from a Magento (Adobe Commerce) merchant we worked with. They were running the classic Luma theme with tons of customizations. Their CWV scores were a disaster – LCP around 4-5s on mobile, FID often 300ms+, CLS not terrible but not great either. After much debate (and hearing me evangelize), they decided to try a radical approach: implement the Hyvä theme as a replacement frontend. Within 2 weeks, we had Hyvä up on their staging environment, replicated their design (with some improvements), and went live. **Immediately, every Core Web Vital turned "good" (green)**. LCP went from

~4s to ~1.8s on mobile, FID (or INP) dropped under 100ms, CLS basically zero. It was night and day. Even more interesting, their development team's productivity shot up – they could make front-end changes much faster because the codebase was simpler. They started deploying UX improvements weekly instead of monthly. Over the next quarter, this site saw an **11% increase in Q/Q revenue**. Some of that was seasonal, sure, but they outpaced their usual seasonal lift. Digging into the analytics, we saw better organic traffic (likely thanks to better CWV and perhaps slightly improved SEO rankings) and a higher conversion rate (likely thanks to the snappier site – users browsed more and had fewer abandons). Plus, their total cost of ownership went down – Hyvä's license and implementation cost was far less than the custom headless build they had been considering (in fact, we estimated about a **40% cost saving in year one** compared to going headless). The moral: sometimes a bold change like adopting a modern theme or framework can pay for itself many times over. If you're on a platform that offers a more performant front-end stack (Magento/Hyvä, or even moving from a heavy custom Shopify theme to a leaner one), it's worth the upfront investment. Not only do you get a faster site (with all the conversion and SEO perks), but your team can iterate faster, and that agility drives further growth. Speed begets speed, in more ways than one.

These case studies each hit on different aspects – third-party scripts, overall front-end optimization, and wholesale platform/theme change – but they share a common thread: faster sites made more money and provided better user experiences. I encourage you to collect your own internal case studies. When you make a perf improvement, document the before/after and the business impact. It's the best way to convince stakeholders that this stuff matters.

Common Anti-Patterns to Avoid

In the journey to a faster site, it's just as important to know what *not* to do. I've seen well-intentioned teams actually hurt their site (or waste a ton of effort) by chasing the wrong goals. So here are some speed and optimization anti-patterns you should watch out for:

Chasing a 100/100 Lighthouse Score as the end-all, be-all. Yes, it feels like a high score on a test, but obsessing over that perfect 100 can lead you astray. I've literally seen a case where a company installed a third-party Shopify app that **faked their site** when it detected a Lighthouse test, just to get a 100 score – complete nonsense! (The app would serve a nearly empty page to Lighthouse, tricking it into a perfect score, while real users saw the normal site). That "solution" did nothing for real users and once discovered, it was a PR embarrassment. The truth is, as we saw with Amazon, you can score 50 and still have a wildly successful site if the real-user metrics are solid. Conversely, you could game your way to 100 and still have users complaining about slowness. So, don't make 100 your North Star. Focus on actual metrics and improvements. Use Lighthouse to guide fixes (the recommendations are useful), but if you're at, say, a 92 and one remaining suggestion is to remove an analytics script that *you truly need for business*, it's okay to not get the 100. Aim for practical speed, not vanity scores. And absolutely avoid any vendor or plugin promising "instant 100 scores" – they're either snake oil or harmful. True performance gains come from real engineering work, not cheats.

Overloading with Pop-ups and Widgets (CLS nightmares). We all understand the marketing need: collect emails, offer discounts, cross-sell, chat with customers… but if your approach is throwing pop-ups, banners, and other dynamic elements that shift the page, you're murdering your CLS (and likely your user's patience). I've seen sites where the moment you try to scroll, a newsletter pop-up appears and shoves the content down – the user's about to click a product, and now they've clicked something else. Not good. High CLS erodes user trust; it makes your site feel janky. As we noted, a big culprit is things like "giant pop-ups" unexpectedly taking over the screen. The anti-pattern here is not the pop-up itself (which might be effective in moderation), but implementing it in a way that *reflows* content. Use overlays or modals that don't push the page layout when they appear (i.e., they appear on top of an overlay). Or, if something like a promo banner must insert above content, at least reserve the space via CSS so that when it loads, it doesn't shove everything down. Better yet, reconsider if you need so many distracting pop-ups. Each one might have a conversion purpose in isolation, but collectively they can slow down and destabilize the experience. The speed-first mindset often leads to *simplicity*: one primary call-to-action instead of five competing widgets.

Leaving the site in Developer Mode (or forgetting cache settings). This is an easy mistake, but I see it surprisingly often, especially when agencies or dev teams rush a launch. Many platforms have a dev mode that disables caching, concatenation, minification – all the performance goodies – to facilitate rapid development. That's fine *in staging*, but if you deploy to production and forget to flip to prod mode, you're essentially shipping a slow version of your site. I've encountered Magento sites left in dev mode after launch – meaning every page load was uncached, every resource unminified, leading to *severely* degraded performance. As one colleague put it, it's like driving a car with the parking brake on. In one instance, simply switching the site to production mode (a one-click admin setting) improved load times by 2-3x because it allowed full-page caching and static content caching to work. Similarly, ensure that CDN or caching plugins are properly configured and *enabled*. An anti-pattern is to do all the hard work of optimizing, and then not reap the benefits because of a misconfiguration. Make a post-deployment checklist that includes verifying caches are warm and working. In summary: don't shoot yourself in the foot with a config oversight.

Going "Headless SPA" just for the sake of it (when a simpler solution suffices). This might be a bit controversial, but it needs to be said. "Headless" architectures (where you decouple the front-end and use a JS framework for the storefront) have been all the rage. They *can* solve certain problems and offer flexibility, but they are **not a guaranteed performance fix**. In fact, if done poorly, a custom headless front-end can be slower than a well-optimized traditional setup. I always caution: don't go headless just because it's trendy or because someone sold it as a silver bullet for speed. It's a heavy solution – you're rebuilding a lot from scratch. Unless you truly need the separate front-end (for example, a unified app experience, or using one front-end for multiple backends), consider options like performance-focused themes or partial headless (hydration, etc.) first. We've had clients itching to scrap their platform's native theme for a React app purely for speed, but we achieved the speed in a fraction of the time by using something like Hyvä (for Magento) or simply refactoring the theme. Hyvä, in particular, is a nice middle ground: you stay within the Magento ecosystem (so all features work out-of-the-box, no compatibility nightmares) but get a modern front-end. It's not magic either, but it avoids the common headless trap of *shifting* complexity around. The anti-pattern here is **over-engineering**: adopting a more complex architecture than needed. If your dev team is all React and really wants to go headless, fine – just ensure you budget for a marathon, not a sprint. But if your goal is to improve speed and you can do that by simplifying (less JS, better caching, maybe a newer theme), do that. In short, don't rebuild the whole house when maybe you just need to renovate a room. As I often say, **Hyvä isn't magic, but neither is headless** – both require work, so choose the approach that gives maximum payoff for effort. For many, the "lightweight theme" route is the unsung hero.

Keep these anti-patterns in mind. Whenever you're working on performance, ask "Am I doing this for a real user benefit, or just to game a metric?" and "Is there a simpler way to achieve this?" If you stay user-focused and simplicity-focused, you'll avoid 90% of pitfalls.

With the mistakes to avoid covered, let's pivot to something fun: how can different types of ecommerce businesses approach site speed? Not every store is the same, so strategies might differ for a B2B vs a D2C brand. I've sketched out a few archetypes with tailored playbooks.

Performance Playbooks by Ecommerce Model

Every ecommerce business has its own quirks and priorities. Here are three common archetypes I've worked with, and a quick playbook for each to maximize site speed returns:

"Side-Car" B2B Distributor – This is the B2B company that might have a primary sales channel via sales reps or EDI, but also runs an ecommerce portal for clients (often on desktop). The site might be heavy with data (lots of SKUs, complex pricing logic, maybe pulling info from an ERP on the fly). *Playbook:* Focus on back-end efficiency and caching. B2B users may tolerate slightly less flashy UI, but they will not tolerate a slow product search or price load. If you have pricing APIs or inventory calls happening during page load, **cache those responses** aggressively. Even if prices must be real-time, consider edge caching them for a few minutes – 99% of the time, they don't change that fast. Aim for an LCP under 3s on desktop for your catalog and product pages as a baseline. These users are often on work computers (maybe not as powerful as latest iPhones, but decent) and possibly on company networks. Interestingly, I advise testing your site on a crappy old Windows laptop with Internet Explorer (yes, some corporates still use IE or old Edge) because a subset of B2B customers will be in that environment. Make sure your site still loads fast there or at least degrades gracefully. Watch your **CLS** too, especially if you have a lot of dynamic content modules. B2B users might be comparing specs and if the page shifts around, it's super annoying. I recall a B2B client where simply pre-setting the height for a technical specifications table (which loaded async) avoided a big layout jump that was frustrating users. Another tip: because B2B sessions can be long (people might leave a cart open, etc.), memory leaks in SPAs can accumulate. If you have a single-page app for B2B, ensure you periodically hard refresh or encourage it. In summary, for B2B: cache aggressively, optimize for stable and *reliable* performance more than fancy animations, and test on typical corporate setups. If you do it right, you'll get that LCP < 3s on desktop and have very happy corporate buyers who can quickly place reorders without hiccup.

Omnichannel Retailer (Brick-and-Click) – This is a retailer with physical stores and an online presence. A common scenario: customers might be browsing on their phones while in a store, or checking product availability on store WiFi. *Playbook:* Mobile-first optimization is key, but also think about weird network conditions. For instance, many store WiFi networks are slow or have firewall rules. I've seen in-store WiFi block certain scripts (e.g., Facebook or Google tags), which can actually hang a site if not loaded with proper timeouts. So, **test your site on a store-like network** – maybe a throttled 3G or a network that simulates high latency. Make sure the core shopping experience works even if some third-party tags fail. Omnichannel also often means lots of promotions and maybe a heavier homepage (since you're showcasing in-store events, etc.). Be careful with **CLS** here: big promo banners that load in, or a carousel of store locations, can cause layout shifts. One recommendation is to use skeleton placeholders or reserved space for any personalized or geolocation content. For example, if your site shows "Your Store: Dallas" and that loads after checking the user's location, allocate the div size in advance so nothing jumps when the info arrives. Another omnichannel consideration: store locator maps. If you have a store locator page with an embedded map, lazy-load that map. It's usually heavy. Only load the map once the user interacts with the "find a store" section. Similarly, user reviews or social feeds – anything pulling from external – lazy-load them after the main content. The idea is an omnichannel retailer's site often tries to do *everything* (product info, branding, community, store info), so prioritize what loads first (product info and images) and let the rest come later. Also, if you have an integrated **POS or inventory check** on the product pages (e.g., "Check in-store availability"), make sure that's async and doesn't block the initial page render. Ultimately, test on a variety of mobile devices (low-end Android especially) and on a slow network – your target is to keep things usable under those conditions. And if possible, measure in-store usage: if you see a lot of on-premise traffic, consider a lightweight version of pages for those contexts. Maybe an AMP-style quick view for products when on store WiFi. Omnichannel customers expect the online and offline to complement each other, and nothing is worse than a store associate telling a customer to check the website, and the site is too slow to load on their phone in the store. Aim for **green CWV on mobile** (so sub-2.5s LCP, etc.) and you'll be ahead of most.

Pure-Play Digital D2C – A digital-native brand, selling direct-to-consumer online only. Often heavy on media and storytelling, maybe built on Shopify or a custom stack, with a lot of mobile traffic via social media referrals. *Playbook:*You live and die by your mobile site speed and smoothness. These users will bounce if things are slow *at all*. So obsess over **mobile CWV**. I'd say push beyond the 2.5s LCP – try to get <2s or even <1s on your landing pages if you can (it *is* possible with careful optimization and CDNs). One trick: use a fast global CDN and even consider geo-routing or edge computing for critical content. Some D2C brands set up service worker caching or use prerendering for the first visit so that when a user clicks an Instagram ad, the landing page loads almost instantly. This can be done by warming edge caches or using something like Cloudflare Workers to generate pages closer to users. It's advanced, but it can give you that sub-1s visceral instant feel. Definitely A/B test these improvements. For instance, try an experiment where 50% of users get a version of the site with an additional edge cache layer for HTML, and compare conversion rates. If you see a lift, you have ammo to invest more in such tech. Another focus: **JavaScript bundle size**. Many digital brands use modern JS frameworks; ensure your bundles are split and as lean as possible for first load. If you can serve initially only the minimal JS/CSS to render the above-fold content, do that and lazy-load the rest. Also, a pure-play needs to watch for **regressions** as you add new fancy integrations. It's tempting to install that chat bot, that webAR plugin, etc. Have a performance budget: e.g., "Homepage LCP must stay under 2s, any new feature that threatens that needs justification." And measure the *impact* of speed on conversion continually – it will likely show strong correlation. I've worked with a D2C brand that discovered every 100ms of load time improvement was netting a ~0.5% increase in add-to-cart rate – so they baked performance into their growth strategy. For pure-plays, I also suggest experimenting with **CDN edge rules** – for example, if you can detect a first-time visitor from an ad, maybe prerender the page for them (some CDNs allow you to tailor caching or responses based on headers). This gets complex, but even simpler: ensure your CDN is caching pages for anonymous users. Many D2C sites are mostly public content except the cart, so leverage that. Aim for blazing fast, feel-it-in-your-bones speed. If you achieve something like <1s LCP and <100ms INP on mobile for first-time visitors, you'll likely blow the socks off new customers and have an edge over competitors. **Speed is part of your brand** in the digital realm – nobody remembers a pretty site that was frustratingly slow, but they will remember a site that felt instant and fluid to shop on.

Of course, these are generalized playbooks. But I hope it gets you thinking about what *specific* strategies suit your business. The main thing is to align your performance efforts with your user context: B2B desktop users differ from millennial mobile shoppers, and you should optimize accordingly.

Now, regardless of what kind of site you run, you're probably wondering: how do I keep track of all these improvements? What metrics should I watch day-to-day or week-to-week to ensure we're on track? In the next section, I'll outline the key metrics I recommend putting on your performance scorecard (in fact, we include these in our EOS – Entrepreneurial Operating System – scorecards for ecommerce teams).

KEY METRICS TO WATCH (PERFORMANCE SCORECARD)

To truly make site speed a part of your growth strategy, you need to track it like you track revenue or conversion rates. Here are the **metrics** I suggest you watch regularly (at least weekly, if not daily). These are typically measured at the 75th percentile (p75) of real-user data unless noted otherwise, because that's how Google evaluates CWV and it filters out the extremes while still capturing slower experiences:

LCP (Largest Contentful Paint) – p75 < 2.5s on mobile. This is the big one for loading. Ideally, you'll track both mobile and desktop, but mobile is usually the laggard. If your p75 LCP is above 2.5s, you know 25% of your users have a sub-par experience. Work to bring it down. Once you're near 2.5s, try to keep improving towards 2.0s or below – but 2.5s is the official "good" cut-off.

INP (Interaction to Next Paint) – p75 < 200ms. This ensures your site is responsive. If p75 INP is, say, 300ms, one in four users experiences noticeable lag. Track this especially after deploying new features or third-party tags – those often impact interactivity. If INP isn't available in your tools yet (it's relatively new), you can use FID as a proxy, but eventually INP is the one to watch.

CLS (Cumulative Layout Shift) – p75 < 0.10. This one's usually easier to keep green, but if you run lots of ads or dynamically injected content, keep an eye on it. Any spike above 0.1 means something caused layout jank for a good chunk of users. Track CLS to quickly spot if a new pop-up or ad integration starts causing issues.

Total Blocking Time (TBT) – Lab metric < 300ms. While TBT isn't directly a CWV, I recommend keeping an eye on it in your synthetic tests (Lighthouse) because it's a good proxy for overall JS bloat and correlates with FID/INP. A good target is < 300ms TBT in Lighthouse tests. If your TBT suddenly jumps (say after adding a new script), it's a red flag that interactivity might suffer. It's easier to watch TBT in a controlled test than INP in the field sometimes, so use both.

Error-Free Sessions Ratio – > 98%. This one is about quality: what percentage of user sessions go through without any uncaught Java-Script errors? If your site throws errors, those can kill performance (a JS error might break other scripts) and certainly hurt UX. Using a tool like New Relic, Datadog, or even TrackJS/Bugsnag, track the stability. I aim for 98%+ of sessions with zero errors. If you deploy something and that drops to 95%, you introduced a bug or conflict that could also slow things down (errors often break optimizations like lazy-loading). High stability correlates with smoother performance. Plus, it's just good practice – speed won't matter if your checkout is broken for 5% of users due to a JS error.

Core Web Vitals "pass rate." Google's definition of passing is all three CWVs good at p75. It's handy to track the percentage of your pages (or overall site) that are "Good" in CWV as per Search Console. It's either pass or fail. If only 70% of your pages pass currently, set a goal to get to 90%+. This is more of an aggregate, but it's literally what Google uses for ranking (either you pass or not). So in SEO reports, include that stat.

Keep these metrics visible. We often add them to a client's KPI dashboard alongside traffic and conversions. When you do your weekly team meeting, and you see LCP p75 went from 2.7s to 3.5s, that prompts a discussion: "What happened? Did we push a heavy change? Did traffic from a new region (with slower connections) skew it?" It creates accountability. Likewise, celebrate improvements: if you hit all-green across the board, high five the dev team.

Alright, we've covered a ton: from the concept of speed as a revenue driver, to metrics, to tools, to case studies. At this point you might be thinking, "This is great, but where do I actually *start* on Monday morning?" Perfect segue – I've got a quick-start plan for you.

THE 7-DAY SPEED RUN: QUICK-WINS SPRINT

Sometimes you just need to roll up your sleeves and knock out a bunch of improvements. Here's a **one-week challenge** I give to teams who want to kickstart their performance journey. Follow this schedule, one focused task per day, and by next week you'll already see a difference:

Monday: Benchmark with PageSpeed Insights. Pick your homepage and a key product page (or category page). Run them through PageSpeed Insights and save the reports. Note the CWV metrics and major recommendations. Also, if you have access to Search Console's Core Web Vitals report, note how many URLs are failing. This is your "before" snapshot. Share it with the team – nothing like a little red/yellow in those charts to motivate everyone.

Tuesday: Eliminate one unused script or heavy plugin. Remember that script audit we talked about? You don't have to do it all at once. Identify at least one third-party script or plugin that is non-essential and remove it (or disable it) as a test. Maybe it's a heatmap tool you haven't checked in ages, or a fancy widget that's not pulling its weight. Remove it and monitor for any issues (there likely won't be if it was truly unused). This could easily save hundreds of milliseconds or more. Quick win.

Wednesday: Compress your top 10 images. Find the 10 largest image files used on your site (your dev or even PageSpeed can help identify these). Use an image optimization tool (there are many online or built into dev pipelines) to compress them without visible quality loss. If you're not already using WebP or AVIF for these, convert them. Replace those images with the optimized versions. This could easily chop off a couple of megabytes, which is huge for load time. Users won't notice any visual difference, but they will notice faster rendering.

Thursday: Enable full-page caching (or equivalent). If you're on a platform like Magento or WordPress and not using a full-page cache, get it turned on *now*. This might be installing a plugin or simply toggling a config. For example, Magento's built-in full-page cache or a Varnish server, or WP's caching plugins like WP Super Cache. For Shopify or other hosted platforms, ensure that caching headers are enabled via theme (Shopify largely handles it for you, but apps can interfere). The goal today is to make sure a returning visitor or a second pageview loads much faster than the first. Full-page cache does that by serving cached HTML. Test it: navigate your site twice and see if the second load is near-instant (it should be if caching is working). If it's not, figure out why. This might require a dev if not straightforward, but it's worth it.

Friday: Add a CDN or edge rule. By Friday, let's tackle the network layer. If you're not behind a CDN, see if you can integrate a free one like Cloudflare (even free Cloudflare will cache static assets and give some edge acceleration). If you already have a CDN, work on an edge rule – for instance, an edge cache for specific pages (like cache your homepage HTML for 30 seconds globally, so bursts of traffic hit the edge, not your server). Or an image optimization rule through the CDN. Essentially, implement one network-level optimization. Another simple one: enable HTTP/2 or HTTP/3 if not already (many CDNs do this by default). This will boost how assets load in parallel. By end of day, you want to proudly say, "Our site is now being delivered through [CDN name]" or "We just enabled HTTP/3 support" etc.

Saturday/Sunday: Set up GA4 Alerts for CWV. Alright, it's the weekend – time for some light work (and maybe some R&R). If possible, use a bit of weekend time to set up a dashboard or alert. In Google Looker Studio (Data Studio), you can create a simple report using the Google Analytics 4 data or Chrome UX Report data to track CWV over time. Configure an **alert**: for instance, if average LCP for the week goes above 3s, email an alert; or if % of good CWV drops below X%. GA4 doesn't natively send alerts for CWV, so you might use a third-party tool or even just schedule a weekly email of the report. The point is to automate visibility. This might take a couple of hours to tinker with, but once it's done, you'll have something in place to keep you informed without manual effort. Consider it an insurance policy on all the work you're doing.

Next Monday: Re-run PageSpeed and celebrate. It's been a week. Time to measure the impact. Run PageSpeed Insights again on the same pages as last Monday. Compare the metrics. Ideally, you see some improvement – maybe LCP went from 4s to 2.8s, or the performance score jumped from 50 to 80. Even if modest, any progress is good. Share the before/after with the team and leadership. Nothing builds momentum like quick wins. If you followed the steps, you likely removed a major slowpoke script and optimized imagery, so I'd expect noticeable gains. And if something didn't budge much, that's instructive too – it means the remaining issues might be deeper (or maybe you were already good in that area). Use this to plan your next moves.

This 7-day sprint is just the beginning, but look at what you did in essentially a few hours spread over a week: you measured baseline, fixed a script, optimized images, improved caching, upgraded network delivery, and set up monitoring. That's an excellent foundation. Rinse and repeat similar cycles and you'll be in the top tier of ecommerce performers in no time.

Finally, let's wrap up by talking about the human element – how to **embed a culture of speed** in your organization, so all these improvements stick and keep coming.

Building a "Speed Culture" in Your Team

Sustainable site speed isn't a one-off project or a checklist you do once. It's a mindset that needs to permeate your team's culture. When speed is part of your DNA, you'll find performance issues being caught and fixed as a matter of course, rather than in big panicky retrofits. Here are a few ways to embed this performance-first culture:

Performance Budgets as Gatekeepers. Just like you wouldn't deploy a feature that fails QA tests, set up **performance budgets** that every new feature or page must meet before it goes live. For example, declare that any new page should keep LCP under 2.5s and add less than 100KB of JS. If a dev opens a pull request that would, say, add a 500KB library, the CI could flag it, or at least the team should discuss it. Many build tools can calculate bundle size diffs; incorporate that into your workflow. It might feel like extra work initially, but it forces trade-off discussions early. I often tell teams: performance is a requirement, not a "nice to have." Treat it like one. That way, people know from the start that shipping something new isn't just about features working, but also about not slowing down the site.

Reward and Recognize Speed Improvements. People respond to incentives and appreciation. Make site speed a visible metric and celebrate wins. Did someone refactor code that shaved 200ms off load time? Shout it out in the team meeting. Maybe do a fun thing like *"Speed Champion of the Month"* award. I've even seen companies buy pizza or team lunch when they hit an all-green Core Web Vitals milestone. These little celebrations reinforce that performance work is valued, not just invisible labor. When the marketing team sees the devs get kudos (and pizza) for performance gains, they'll understand why sometimes their tag requests are challenged. It becomes a team victory. At Creatuity, we literally cheer when a client's site moves into the green across the board, because we know it's going to positively impact their business and make our lives easier too (fewer fire drills).

Bake Performance into QA and Testing. Make load times and performance metrics part of your testing protocol. For instance, after every deployment, run automated Lighthouse tests on key pages and compare to baseline. If something regresses by more than X, consider rolling back or fixing forward immediately. Also include performance in your QA checklist: "Does the page load within 3 seconds on simulated 4G? Yes/No." When QA flags it, it becomes a blocker just like a functional bug. This process ensures you catch issues early. It also sends a message to everyone – from product managers to designers – that speed isn't an afterthought.

Quarterly "Chaos" Load Tests. This is a bit like fire drills. Once a quarter (or before major peak events), run load tests and simulate high traffic to see how your site and infrastructure hold up. Involve the whole team in reviewing the results. These load tests might uncover slow database queries, memory leaks, or other issues that only appear under stress. By scheduling them regularly, you normalize the practice of performance tuning as part of maintenance. It's much better to find out in a controlled test that your site slows down at 500 concurrent users than to discover it on Black Friday. And when you fix those issues, again, celebrate. It's like practicing for a sports event – when the real thing comes, you're ready.

Cross-functional Buy-in. A culture of speed isn't just a dev team thing. Educate your marketing and creative folks on why, for example, a huge 10MB video background might hurt conversions, or why loading five different tag managers is a bad idea. When non-technical team members understand the *why*, they often become allies. I've seen marketing teams proactively ask "Will this campaign landing page still load fast with these assets?" – that's when you know the culture shift has happened. One idea is to include a quick performance intro in onboarding new team members, so everyone who joins knows it's part of your company's values.

Ultimately, the goal is to reach a point where performance is considered at every stage: from design ("Can we make this animation lighter?") to development ("Can we reuse an existing library instead of adding a new one?") to infrastructure ("Do we have enough server capacity in region X for the sale?"). It's just like how security is now everyone's job – performance needs that treatment too in modern ecom teams.

One more thing: instill the mindset that **site speed is never "done."** It's like fitness – you have to maintain it. The web keeps evolving, browsers change, new scripts get added. But if you have a culture of continuously watching and tweaking, you'll keep in great shape.

Alright, we've covered a ton of ground in this chapter. From vivid tales of failure to detailed metrics and tooling, to team culture. By now, you should be convinced that site speed and Core Web Vitals truly are a revenue multiplier, and you have a playbook to go make it happen.

As we wrap up, I want to emphasize: **site speed is one of those rare win-wins**. Your customers get a better experience, and you get more conversions and higher SEO. There are not many levers where improving something for the user *directly* improves business metrics in such a linear way. That's why I'm so passionate about it.

Keep this chapter handy as you implement changes, and remember – every millisecond counts (and adds up). Now go forth and make the web faster!

Next up: in Chapter 4.1, we'll dive into Org Design for E-commerce 3.0, where I'll talk about how to structure teams and processes to continuously ship improvements like these. Spoiler: it involves breaking silos between dev, design, and marketing so that speed (and other key goals) are owned by everyone. After all, having a "speed culture" often starts with having the right team structure. Stay tuned for that, and in the meantime, enjoy the sweet victory of faster load times!

4.1: ORG DESIGN FOR E-COMMERCE 3.0

I walked into a **legacy retail company** a couple of years ago, and what I saw was like stepping into a time machine. Their e-commerce "team" consisted of a few people tucked away in a corner, completely siloed from the rest of the business. They had to beg IT for every little website update. Marketing treated the web store as an afterthought. The store managers? They saw online sales as competition, not collaboration. The result was predictable: endless delays, finger-pointing, and missed opportunities. It felt like they were running e-commerce like it was still 1999 – or at best 2015 – and in today's market, that's a recipe for losing money **before you even write a line of code**. The shock was that this wasn't *1999*. This was a modern retailer, yet their org chart and mindset were stuck in the past.

That cold dose of reality crystalized something for me: **how you structure your e-commerce team can make or break your growth**. In the early days of online retail, you could get away with a small, isolated web department. Not anymore. Things have changed – customers have **more choices and more competition is coming online every day**. If you're still treating e-commerce as a side project, you're going to get left behind. *If you're still running your e-commerce team like it's 2015, you're probably losing money and don't even realize it.* The legacy retailer in my story learned this the hard way. Fortunately, you don't have to. In this chapter, I'll show you how to **redesign your organization for the e-commerce 3.0 era**, so you can break down silos, move fast, and start shipping results.

WHY THIS MATTERS NOW

Welcome to **E-commerce 3.0** – a world where online and offline blur, and where mid-market players can't afford slow launches or turf wars. Today's customers demand seamless experiences, and new competitors pop up overnight. To stand out, you need both **innovation and flawless execution**. That means your team structure must enable speed, collaboration, and accountability. It's not just about having the right people – it's about organizing them the right way.

Think about your own company: Do you have a clear owner for your e-commerce P&L? Can your team deploy a site update or campaign in days, or does every change require five approvals and a Ouija board to schedule? I've seen too many mid-market firms where e-commerce is treated as a **sideshow to the main business – maybe contributing 10% of revenue – and stuck under some other department's shadow**. At B2B companies this shows up as a culture where the **sales team calls the shots and e-commerce is just along for the ride**. E-commerce staff spend their time fighting internal battles for resources instead of fighting for customers And omnichannel retailers often get caught in internal tug-of-war: drive more online sales and the stores complain you're stealing their lunch. I've literally mediated meetings where store managers were furious about fulfilling online orders, until we fixed the incentives and attribution. (When we properly **attributed foot traffic and sales boosts from Buy Online Pick Up In Store to those local stores, suddenly those managers became e-commerce's biggest fans**.)

The stakes are high. As one new e-commerce director wisely did, you should align your goals with the CEO's goals from day one. If top leadership wants, say, $50 million in online revenue next year, but they're only budgeting $50k for the e-com team, there's a massive disconnect. A *modern e-commerce org design* forces those conversations into the open and gets everyone on the same page. It ensures you have **the right team, with the right mandate, to hit ambitious goals** –

instead of setting goals in a vacuum and hoping some overworked "web person" magically achieves them.

So *why does this matter now*? Because in today's hyper-competitive environment, **speed, agility, and alignment** are your secret weapons. The old silos and outdated mental models are slowing you down. It's time to upgrade how we organize our people. Over the past two decades, through founding an e-commerce agency and advising 500+ online businesses, I've seen what works and what spectacularly doesn't. I even learned the hard way in my own company, Creatuity: early on, *we were all developers doing everything – coding, billing, selling, project managing – which was a train wreck waiting to happen*. We survived about three years like that before an intervention – colleagues basically shaking me and saying "Josh, you **need more than just developers** on your team!". I finally brought in project managers, QA, account managers – the *balance* we needed. The experience ingrained in me that **team structure is not a luxury; it's foundational.**

Now, let's talk solutions. I'm going to introduce a new way to structure your e-commerce organization – a three-layer "pod" structure – and show you how to customize it for *your* business (whether you're a Sidecar, an Omnichannel, or a Pure-Play). We'll cover practical tools to make it work (RACI charts, KPI dashboards, reorg sprints, etc.), and I'll share some field stories (the good, the bad, and the ugly) from real companies who made the leap. By the end of this chapter, you should have a clear blueprint for **Org Design 3.0** that you can start implementing tomorrow. Let's dive in.

The Three-Layer E-commerce Team

Every successful e-commerce operation I've seen, no matter the industry, has a **universal core structure** – think of it as a three-layer engine that drives growth. Titles vary, and one person might wear multiple hats in smaller teams, but the functions remain surprisingly consistent. I break this down into **three groups (or "pods")**: the **Growth Pod**, the **Platform Team**, and the **Enablement Team**. In my view, **every e-commerce business needs someone covering each of these three areas**, whether they're in-house or outsourced. Let's unpack each layer and the roles involved.

The Growth Pod: Your Revenue Factory

The Growth Pod is the **front-line revenue engine** of your e-commerce org. This is where new e-commerce sales are generated, experiments run, and customer experience optimized. In a mature team, the Growth Pod typically includes five key roles:

Merchandising Lead – This person owns the product catalog and online merchandising strategy. They decide *what* you sell online, for how much, with what promotions. In short, they ensure you're offering the right products at the right price to make customers click "Buy". (On small teams, this might be a category manager or even the marketing manager doubling as the online merchandiser.)

User Experience (UX) Designer – Someone has to make sure the site isn't an ugly, confusing mess. The UX lead designs and optimizes the on-site experience so that it's pleasing and friction-free for the customer. From layout to navigation to mobile responsiveness, they champion the customer's ease of use. You'd be amazed how often this role is absent in legacy teams – and how much money gets left on the table due to poor design.

Data Analyst – I often call this the secret weapon of the Growth Pod. A dedicated analyst will **dig into your data and turn it into action**. They clean up messy analytics, build dashboards, and surface insights. A good analyst can tell you, "Hey, mobile conversion is lagging on category X" or "Customers from channel Y tend to churn after 2 orders." They'll even propose A/B tests based on trends they spot, and crucially, they'll measure the results of those experiments. If you're not a data-driven e-com team yet, hire or designate a data analyst *yesterday*.

Developers (Front-end & Back-end) – No surprise here: you need folks who actually build and maintain the online store. I generally recommend having at least one front-end developer (for the user inter-face, site speed, HTML/CSS/JS stuff) and one back-end developer (for platform code, integrations, server-side logic) – or a couple of full-stack developers – within the Growth Pod. These are the builders turning the ideas into reality, whether it's launching a new feature or just keeping the site live on Cyber Monday. They'll be the ones executing deployments, so you want them closely aligned with the merch, UX, and analyst roles day-to-day.

Product Owner – This role is the **single biggest game-changer** for most teams, and unfortunately the most often missing. The Product Owner is a **senior business leader** who owns the e-commerce **vision and P&L** and prioritizes what the team works on. Think of them as the mini-CEO of the online channel. They decide, "We're focusing on improving search this quarter, not launching a new catalog, because it will move the needle more." They are empowered to make call-after-call about scope, priorities, and what "success" looks like for the website. When companies lack a true Product Owner, projects derail and teams flounder – I see it *all the time*. (One of the **worst mistakes** is to assume a *project manager* can double as a Product Owner – or to force one person to do both jobs. Please don't do that; I'll explain why in a bit.)

In a perfect world, your Growth Pod is a **dedicated, multidiscipli-nary squad** focused on growing online revenue. They operate almost like a startup within your company – fast, cross-functional, and laser-focused on conversion and sales. If you're a smaller outfit, you might not have five separate humans for these roles; that's okay. Maybe your VP of Marketing is acting as Product Owner *and* Merchandising Lead, while your lone web developer wears the front-end and back-end hats. Maybe you're outsourcing UX design to an agency. Do what you must – but **make sure every one of these functions is owned by someone**, internally or externally. Too often I see a "team" of a few developers and a marketing manager trying to do e-commerce, with nobody thinking about UX or owning the roadmap. That won't cut it anymore.

One more note: **Product Owner vs. Project Manager**. These are two distinct roles and combining them is a recipe for conflict. The Product Owner cares about *what* and *why* – what features or initiatives will drive value, why we prioritize A over B. The Project Manager (if you have one) cares about *how* and *when* – how we execute on time and on budget, when each task is delivered. Those focuses can clash: a Product Owner might say "We need to add a subscription feature for long-term value," while a Project Manager might groan about it delaying the timeline. If the same person is trying to fill both shoes, either the vision or the delivery will suffer. I advise separating those concerns. Many mid-market teams don't even have a formal project manager, and that's fine – the agency or developers often handle task management. But *do*have a clear Product Owner. **Empower that person** to make decisions. It will speed everything up: for example, when my agency works with a client who has a true Product Owner, we can ask one question and get an immediate answer, instead of a week-long game of telephone through five people. Projects that could drag on for months start launching in weeks.

The Platform Team: Stability and Scale

Next, let's talk about the **Platform Team**. These folks are the guardians of your technology stack – they keep the site fast, secure, and scalable. Now, depending on your business, you **might not need much of a Platform Team in-house** at all. If you're on a fully hosted SaaS platform like Shopify or BigCommerce, or a PaaS like Adobe Commerce Cloud, a lot of the "platform" heavy lifting is handled by the vendor. In those cases, your "Platform Team" might be an external agency or just a part-time role. However, if you run on a more custom or self-hosted setup (Magento Open Source, custom headless implementations, etc.), then these roles become vital internally.

In a classic setup, the Platform Team has two main roles:

Systems Architect – This person designs and oversees the **technical architecture** of your e-commerce ecosystem. They choose the tech stack and ensure all the pieces (commerce platform, CMS, middleware, integrations, servers, networks, etc.) fit together and meet the business needs. They are responsible for stability, performance, and scalability at a high level. In many companies, this is an experienced IT engineer or an enterprise architect who gets "dotted-lined" into the e-commerce group. Even if it's not their full-time job, someone needs to be thinking about how all the systems work together and how to avoid outages when traffic spikes.

DevOps Engineer – The DevOps role is all about **automation, deployments, and infrastructure management**. They make sure your development pipeline runs smoothly – code can be tested, deployed, and rolled back if needed, with as much automation as possible. They also often handle cloud infrastructure scripts, monitoring, and security updates. If "Systems Architect" is the city planner, "DevOps Engineer" is the master plumber/electrician making sure the utilities run reliably.

Often, in smaller teams, a generalist IT person or an outside agency covers this. I've seen plenty of mid-market retailers where the same IT admin who manages corporate email is also the de facto DevOps for e-commerce – which is fine if it works. Just ensure it's clear *who* is on the hook for keeping your site up at 2 AM on Black Friday. If the answer is "uh, nobody," then you have a gap to fill.

For many **Sidecar B2B** businesses where e-commerce is small, the Platform Team might not be a dedicated team at all – it could be just *a few hours of an IT staffer's week* to keep the lights on. And that's okay to start with. But as you grow, don't neglect these functions. If you plan to innovate with new tech (say, headless front-ends, PWAs, fancy search engines), having a strong architect and DevOps capability will save you a lot of pain. The key is to **map out these responsibilities explicitly**. Too many times I ask "Who's responsible for site uptime and deployments?" and get blank stares or five people pointing fingers at each other. If you're on SaaS, make it clear that your platform vendor and maybe your agency cover most of this – but have internal oversight. If you're on a complex stack, invest in at least one go-to technical mastermind internally or on retainer.

THE ENABLEMENT TEAM: MAKING EVERYONE SMARTER AND HAPPIER

The third layer is one most companies haven't formally recognized (yet) – the **Enablement Team**. These roles don't directly drive revenue or code the site, but they *enable* the rest of the org to succeed in e-commerce. This team makes the other teams **smarter, faster, and more customer-centric**. It's all about support functions that ensure your Growth Pod and Platform Team are working on the right things and that the *organization* can absorb the changes e-commerce brings.

Here are the key roles I slot into Enablement:

Analytics/Attribution Specialist – You might call this a data strategist or business analyst; the point is, someone needs to own **analytics strategy and revenue attribution**. Why? Because when e-commerce is growing, it often steps on the toes of other channels. Stores and sales reps will squabble over who gets credit for a sale. I've heard a sales guy claim, "Well I talked to that customer last year, so I deserve credit for *all* their online reorders," while the e-com team says, "Hey, the customer is clearly reordering through our website's quick-order tool with no sales involvement, so that revenue should count as e-com." These attribution battles can turn ugly and derail your progress. The solution is to assign an independent Analytics Specialist to set the rules and analyze performance objectively. This person tracks overall KPIs, builds the attribution models, and might manage your Google Analytics, BI dashboards, etc. Crucially, **don't bury this under marketing or sales** – put it in the Enablement team so they can arbitrate fairly. (In a small org, your Data Analyst from the Growth Pod might double up on this role , or you might engage a third-party analytics expert. Just ensure someone is minding the analytics store.)

Customer Experience (CX) Coach – This role is all about **ensuring great customer service and experience** across channels. The CX Coach (or whatever title you choose) will monitor customer feedback, review customer service logs, and make sure the *experience* we're delivering matches our brand promise. For example, they might analyze customer chat transcripts to spot common pain points, or ensure that policies for returns are customer-friendly and consistent online vs in-store. Often this role is not a new hire but an enlightened **customer service manager** who is given a seat at the e-commerce table. I've had CX folks in my teams who became the voice of the customer internally – pointing out, "Hey, our new checkout flow might confuse people who call in to order, let's train the call center on it" or "Our online product descriptions aren't addressing questions that customers later phone us about." They help plug those gaps. Include them in your projects and meetings; it pays off.

Training Coordinator (Internal Training & Change Management) – Whenever you launch new e-commerce tools or features, someone has to train the staff (and sometimes even educate customers). A Training Coordinator makes sure **new processes and platforms are rolled out smoothly**. Say you implement a new Order Management System or add an in-store pickup process – this person plans the training for store associates, the cheat sheets for customer service, maybe even the customer-facing FAQs. They think about the human side of tech changes. I recently spoke with a company about to re-platform their entire site; my advice was that their *biggest* challenge wasn't the tech itself, but training their team and users on the new system. That's true for any significant change. The Training Coordinator role can be part of HR or ops in many companies – just ensure they're plugged into e-commerce changes. They might even coordinate training content for customers, especially in B2B scenarios where buyers need to learn new online tools. Don't underestimate this function – a smooth internal rollout can make the difference between a successful project and one that flops because nobody knew how to use the new thing.

Your Enablement Team might not meet every day like the Growth Pod, but they are **critical allies**. They provide the dashboards, the customer insights, the staff training, and the policy guardrails that let your front-line folks move fast without breaking the business. In practice, these roles often come from existing departments – for instance, your current Analytics Manager or your Customer Service Director can be given these charters **as part of the e-commerce Enablement Team**. Explicitly call that out. Let them know, "When we're working on e-commerce initiatives, you're our Analytics Owner," etc. That clarity not only gives them ownership, it also shifts their mindset. Instead of "Ugh, extra work from the web team," they start thinking like part of the e-commerce engine. I've seen a **huge mindset shift** when companies do this: suddenly everyone knows their role in pushing e-com sales, and they collaborate much better.

To summarize the three layers: **Growth Pod drives revenue, Platform Team provides stability, Enablement Team amplifies and aligns the efforts**. When you structure your org this way, you break the old pattern of one isolated "webmaster" or an IT-led project team that doesn't talk to marketing. You end up with a dedicated cross-functional unit that can innovate quickly *and* a support system to integrate that unit with the rest of the company. It's the best of both worlds – autonomy with alignment.

Breaking Down Silos and Outdated Models

Let's confront those **outdated mental models** head-on. The legacy approach (still alarmingly common) is to treat e-commerce as a bolt-on. Maybe it lives under IT, maybe under Marketing, or maybe it's just one enthusiastic manager with a couple of developers off in the corner. In these setups, **silos and bottlenecks abound**. Marketing runs campaigns without telling the web devs. The web devs build features without input from customer service. Store operations refuse to devote resources to online initiatives. It's a mess, and customers feel it.

One client I recall (a mid-sized omnichannel retailer) had an all-too-familiar issue: their e-commerce "team" was basically one *marketing person* and an external dev agency. When that marketing manager wanted to improve the checkout process, she had to go through the internal IT hierarchy – her request went to an IT project manager, who went to the CIO for approval, who then pinged the external developers, who then came back with questions that had to be routed through a QA team… you get the picture. By the time a simple change was implemented, months passed and everyone was frustrated. **Too many handoffs, too many cooks in the kitchen.** We drew a flowchart and counted something like six different handoff points for a basic website update. That's insane – *and expensive*. Every handoff is an opportunity for delays or mistakes (or bugs).

Here's a rule I live by: **no more than three handoffs for any e-commerce task**. I call it the "Rule of Three." If an idea has to bounce between four, five, six people or departments before it goes live, your process is broken. You're probably losing money as a result. In a healthy pod structure, you can often get it down to **two handoffs**: e.g. Merchandising Lead has an idea, Product Owner approves it, developer builds and deploys it – done. Contrast that with the legacy scenario I described: Marketing proposes, IT manager approves, external devs build, QA tests, IT deploys, etc., plus multiple feedback loops. That's why old-school teams feel so slow. *Simplify the communication lines!* When you empower a small cross-functional team, you eliminate those needless loops.

Another outdated mindset is the notion that e-commerce should report into some other department and just "take orders." For example, traditional retailers often tucked e-commerce under Marketing or under Store Operations. The result? E-com never had a seat at the leadership table and couldn't challenge the status quo. We have to **challenge those silos**. E-commerce is a **core revenue center and a strategic asset**, not just a support function. It needs its own voice (hence the Product Owner or a Director of E-commerce who's empowered).

If you're a leader reading this, here's my challenge: **map out your current org chart and project workflow**. Do it visually. Draw who has to talk to whom for decisions. Then circle how many times a project or decision changes hands. Be honest – is it one or two, or is it like a pinball machine? If it's the latter, time to break some glass and reorganize. Sometimes that means a bold move like pulling developers out of IT and into the e-commerce team, or assigning a sales manager to be the e-com Product Owner because they get the customer. It definitely means clarifying accountability. One tool that helps bust silos is a **RACI chart**, which we'll get into next – it forces you to assign exactly one person as accountable for each task, so you can't have the "not my job" syndrome.

Finally, let's debunk the fear that reorganizing is too disruptive. Yes, change is hard – but *not* changing is fatal. The market is evolving whether you like it or not. I love the quote, "Change is inevitable; it's going to happen whether you participate or not." You don't want to be the leader clinging to an outdated org when your competitors have product teams shipping updates weekly. **Mid-market companies actually have an advantage here**: you're not as set in stone as the billion-dollar giants, you can still be agile. I've seen mid-market teams leapfrog larger competitors simply by reorganizing for speed and customer focus. It starts with breaking the old silos and making it crystal clear who owns what in the new model.

Tools and Tactics: RACI, Scorecards, and Sprints

Redesigning your team structure might sound abstract, so let's get very practical. Here are some **tools and tactics** I recommend to implement Org Design 3.0 in a methodical, measurable way.

RACI Chart – Clarify Who Does What: If you're not familiar with RACI, it stands for **Responsible, Accountable, Consulted, Informed**. It's basically a one-page matrix that lists key tasks or decision areas on one axis and team members on the other, with letters indicating each person's role on that task. Use R for the doer (responsible), A for the one **Accountable** (the one neck on the line, must be one person only), C for those you consult for input, and I for those you keep informed.

For example, say we're redesigning the homepage: the Product Owner might be Accountable (A) for the outcome, a UX Designer and Developer are Responsible (R) for executing the work, maybe the Systems Architect is Consulted (C) about performance implications, and the Customer Experience Coach is Informed (I) so support teams know what's coming. By sitting down and mapping a RACI for your major processes (e.g. "Add a new product to site" or "Deploy a code change" or "Plan a marketing campaign"), you **force clarity**. You might discover, "Oh wow, we listed two Accountables for this – no wonder there's tension." Fix that; one task, one throat to choke (Accountable). We did this with the retailer I mentioned earlier, and it immediately exposed overlaps and gaps. It also gave everyone a reference: next time an issue comes up, you don't have five VPs dithering in a meeting – you know who's accountable and who's just consulted. This drastically **reduces finger-pointing** and speeds up decisions.

Handoff Audit – Map the Process: Earlier I talked about drawing your workflow. I highly recommend an exercise we call a **Handoff Audit**. Get your team together in front of a whiteboard (or Zoom) and diagram the steps for a recent project or a typical task, *including every handoff* of responsibility. Then apply the Rule of Three: if there are more than 3 handoffs, ask "How can we eliminate or combine steps?" Sometimes the answer is organizational (e.g. bring QA into the Growth Pod so testing is part of the same team, not a separate department). Sometimes it's process (e.g. allow the developer to push live without waiting for a weekly CAB meeting). One company I worked with discovered that a simple content update went through **four approvals** just because of legacy habit. We collapsed that to one approval in the Growth Pod, and nothing broke – in fact, quality *improved* because the people approving actually knew what they were looking at. This audit becomes your action list for streamlining. It's visual, a bit shocking at times, but incredibly useful.

KPI Dashboards and EOS Scorecards – Measure What Matters: As you roll out a new team structure, you want to keep an eye on whether it's delivering results. I'm a fan of establishing a simple **e-commerce Scorecard** (a concept from the Entrepreneurial Operating System, credit to **EOS Worldwide**). This is a weekly snapshot of key metrics that tell you how the team is performing. Typical metrics might include: deployment frequency (e.g. code releases per week), conversion rate (CVR), average order value, site uptime, page load speed, customer satisfaction scores, and so on – a mix of technical and business KPIs. The Scorecard shouldn't be a data overload; pick 5-10 metrics that you'll watch like a hawk. For example, you might set a benchmark like "at least 1 deployment per week" if you've been historically slow. If the team structure change is working, you should see leading indicators like more frequent deployments, faster cycle times on projects, and gradually improvements in CVR or basket size. Visualize these on a dashboard that the whole team can see. It creates accountability and also celebration moments: "Hey, CVR jumped by 0.2 points after our checkout redesign – kudos team!" Tie this into your regular meetings (next point). The key is to **make success measurable**. As the saying goes, "What gets measured gets managed."

Level 10 Meetings (with Pods) – Keep Alignment Tight: Another EOS practice I've adopted is the weekly **Level 10 meeting** (often called an "L10"). This is a high-efficiency team meeting model (from EOS Worldwide) where you review your Scorecard, discuss rock (project) status, and then dedicate the bulk of time to identifying and solving issues. I won't dive deep into EOS here, but I'll steal shamelessly from it: implement a weekly (or biweekly) e-commerce leadership meeting that includes your Product Owner, key Growth Pod members, and reps from Platform and Enablement (as needed). Keep it to 90 minutes, max. In that meeting, review your metrics, outline any big wins or roadblocks, and then **really discuss the issues** – no agenda fluff. This habit forces the cross-pod communication in a structured way. It also gives a forum to call out, "Hey, we need IT support on X" or "Marketing's campaign is tanking site performance, how do we fix it?" in a collaborative setting. It's amazing how many problems get solved in minutes when you have the right folks in the room weekly – problems that otherwise festered in siloed email threads for months. A client of mine instituted L10-style meetings with her e-com team and saw immediate improvement in decision speed. As one team member quipped, "Conversations that used to take weeks of back-and-forth emails now happen in 5 minutes". Exactly. Meetings (when done right) can actually save time by eliminating confusion. So schedule that cadence and stick to it.

Reorg as a Sprint – Plan, Execute, and Iterate: Finally, treat your **reorganization like a project** – because it is one. Don't just announce "New org starts Monday, good luck!" Rather, run it as a short sprint with clear steps and milestones. For example:

Week 1: Plan and Communicate – Map out the new team structure (who's in Growth Pod, Platform, Enablement). Identify any immediate hires or role changes needed. Announce the vision to the team (explain why we're doing this, what success looks like).

Week 2: RACI and Handoff Audit – Conduct the RACI exercise with the team for key processes , and do the handoff mapping. This not only educates everyone on the new model but also surfaces concerns. Adjust roles or processes based on what you learn.

Week 3: Implement Quick Wins – Pick one or two pilot projects to run with the new pod structure and governance. Maybe it's a small site feature or a marketing campaign. Let the pods operate with their new autonomy and see how it goes. Watch for bottlenecks. This is your chance to iron out kinks.

Week 4: Review and Optimize – Hold a retrospective (retro) meeting. What went well in the pilot? Where did old habits kick in? Perhaps the Product Owner was still going to the CEO for every minor decision (old habit) – coach them to trust their judgment. Maybe developers still weren't looped into planning early enough – fix that. Use this time to tweak the org model.

Week 5-6: Solidify and Expand – By now, you should be seeing improvements (maybe faster deployment on the pilot, or clearer accountability on tasks). Roll the structure out to all e-commerce initiatives. Set up the ongoing Scorecard and weekly meeting rhythm. Also, address any staffing gaps: if it became clear you desperately need a dedicated UX Designer or a DevOps contractor, start that hiring or outsourcing process.

By Week 8 (roughly): you're in full swing with the new org design. From here, treat it as continuous improvement. Keep an eye on those metrics and team health. You might do a follow-up survey or team health check a few months in to ensure the culture is adjusting positively (I highly recommend tools like Patrick Lencioni's team assessment to gauge trust and cohesion – a healthy culture is the fuel for any org structure).

By sprinting through the reorg in a focused way, you avoid the pitfall of "big reorg announcement, then drift." Instead, you execute quickly, demonstrate early wins, and solidify the changes through practice. It also shows your team that this isn't just a flavor-of-the-month idea – you're willing to put in the work and invest in making it stick. Encourage the team to give feedback throughout; sometimes the people on the ground will suggest even better ways to organize or clarify roles.

ADAPTING THE BLUEPRINT TO YOUR TYPE OF BUSINESS

One size does **not** fit all. I've presented the ideal scenario for an e-commerce team, but you absolutely need to **tailor it to your business type**. Earlier, I mentioned three archetypes I use to bucket most companies: *Sidecar B2B*, *Omnichannel Retailer*, and *Pure-Play* (which could be D2C or B2B). Let's discuss how Org Design 3.0 flexes for each of these.

Sidecar B2B: This is a B2B business where e-commerce is maybe 10-20% of revenue or less. The majority comes from traditional channels like a sales force, distributors, or phone orders. E-commerce here is often seen as "sales support". The challenge in sidecar situations is getting the attention and resources for e-com when it's not the primary revenue driver (yet). The three-layer structure still applies, but your **team will be smaller and people wear multiple hats**. You might not have a full Growth *Pod* – it could be one e-com manager (acting as Product Owner + Merchandiser), one shared web developer, and maybe some marketing support. And that's okay to start. What's critical is that you **plan for growth**. Treat e-commerce as a skunkworks that can prove its value. Use consultants or agencies to fill gaps (like UX or DevOps) until you can justify full-time roles. In the Enablement layer, attribution is super important here, because you'll have internal battles with the sales team over credit. I often advise Sidecar e-com leaders to set up a fair **attribution model** (e.g. split credit for online-influenced sales) and get an impartial analytics person or tool to enforce it. This reduces the turf war and helps everyone see e-commerce as growing the pie, not stealing slices. Also, leverage your sales team's knowledge: maybe include a sales rep as a consultant in your Growth Pod for certain decisions (their insight can be gold, say, on what customers are asking for). As e-commerce gains traction (and it will, if structured right), keep revisiting the org – today's sidecar can become tomorrow's main engine. I've seen manufacturers start with e-com as 5% of sales and, a few years later, it's 30% and they're rebranding themselves as a tech-forward distributor. **Plan for that day now**. Start building the pods in embryo form so scaling up is easier when the time comes.

Omnichannel Retailer: You have physical stores or dealers *and* an online channel. Maybe online is anywhere from 5% to 50% of total sales. The key complexity here is the **interplay between e-commerce and brick-and-mortar**. If poorly structured, these teams will compete or, worse, undermine each other. If well structured, they'll feed each other's success. Org Design 3.0 for omnichannel means you **must integrate store ops into your e-commerce strategy**. For instance, your Product Owner should have a strong relationship with whoever heads store operations. They need to jointly solve problems like store fulfillment for online orders, store-exclusive promotions, etc. In terms of team, an omnichannel retailer might have a fully fleshed-out Growth Pod, but the *Enablement Team* is where you often tweak things. The Attribution Specialist here works out how to credit online vs. store sales in blended journeys. The CX Coach might come from the retail side (because they need to ensure store associates deliver on omnichannel promises). And **training** is huge: every time you roll out an omni feature (like BOPIS or curbside returns), you have to train store staff and perhaps customers. From my experience with a retailer called Rural King (100+ stores plus e-com), we saw that when we **launched Buy Online Pick Up In Store, it actually increased foot traffic to every single store** – a win-win, but only possible because we got store managers on board by showing them the benefits. That required org alignment: we had to credit those sales in a way that store managers got **excited** (they saw BOPIS orders as boosting their numbers, not cannibalizing them). So, for omnichannel, **structure your incentives and communication** channels accordingly. Maybe your e-com Product Owner holds a weekly sync with the Head of Stores (to review cross-channel metrics). Maybe you include a store manager representative in major e-com project meetings (so the store perspective is heard early). Organizationally, I've seen success when e-commerce is its own department but very tightly coupled to marketing and stores through dotted lines and committees. The three-layer model still holds, just remember to adapt the **culture**: it's one company serving the customer through multiple channels. Break any lingering "us vs. them" mentality between online and store teams – sometimes a bit of cross-pollination (swap a store ops person into an e-com role or vice versa for a stint) can help everyone appreciate each other's challenges.

Pure-Play Digital (D2C or B2B): In a pure-play, **e-commerce is the business**. This might be a DNVB (digitally native brand) selling direct-to-consumer, or increasingly, a B2B distributor that operates 100% online. In these companies, often e-commerce is already in the DNA – I've met founders of pure-play B2B firms who effectively acted as the e-commerce Product Owner themselves, because that's the core of their business. The advantage here is you usually don't have to fight for a seat at the table; e-com has the spotlight. The challenge is scaling smart and not losing the digital focus as you grow. The three-layer structure for pure-plays tends to be embraced early – these companies know they need merch, marketing, dev, data, etc. all working tightly. You might find less resistance to new roles. However, a couple of watch-outs: Pure-plays can sometimes lack *internal* checks and balances because there is no "other department" to push back. So ensure your Enablement team is empowered to question strategies (e.g. the CX Coach should be unafraid to say "we're getting lots of customer complaints about X, we need to fix that, not just chase new sales"). Also, as pure-plays grow, **don't skip the Platform/DevOps investment**. Many start as scrappy tech teams but at some point you need more formal architecture to handle scale. I've seen hyper-growth D2C brands suffer downtime and slowdown because they outgrew their initial platform and didn't have a strong architect to plan the next step. Get those pieces in place proactively. One more tip: if you're in a **subscription or consumables business (common for pure-plays)**, consider tweaking the Growth Pod by adding a **Retention or**

Subscription Specialist. For example, one of my clients built a subscription service and we embedded a **Churn Analyst** whose job was to monitor subscription KPIs and work with the Product Owner on strategies to improve lifetime value. That might mean adjusting the cadence of renewal offers or adding loyalty perks – it's a bit of a niche role, but it paid off big in retention. Likewise, if you're heavy on **marketplace sales (Amazon, eBay, etc.)**, you might slot a **Marketplace Manager** into the Growth Pod to optimize those channels. Pure-plays often diversify into marketplaces, and that role prevents marketplace efforts from cannibalizing D2C focus (or vice versa).

The bottom line: **know which archetype you resemble and adjust accordingly**. If you're reading this and you have a hybrid model (say, some stores, some wholesale, some D2C e-com), you might pick and choose elements from each approach. The beauty of the three-layer framework is that it's modular. In a Sidecar, your "Platform Team" might be 0.2 of a person. In a Pure-Play, your Platform Team might be an entire department of cloud engineers. In an Omnichannel, your Enablement team spends extra time on cross-channel training. Use the framework as a starting point, but make it yours.

One more field story here: We worked with **Worthington Direct**, a B2B company selling school and office furniture. They started very much as a Sidecar B2B – a traditional sales-driven outfit that gradually built an e-commerce arm. When they finally decided to **re-platform and modernize their e-com**, we helped them put together a more formal e-commerce team structure. It wasn't huge – a handful of people – but we clearly defined the roles: a product owner (who was actually their VP doubling as e-com owner), a marketing/merchandising lead, some developers from our agency side, and an internal IT contact for the platform side. We also set them up with better analytics (Enablement). Post-reorg, they went from infrequent site changes and a clunky old platform to a nimble team that could deploy improvements continuously. The result? Within the first 6 months, their **conversion rate jumped noticeably (double-digit lift)** and they achieved **zero downtime** during a peak-season launch (a big change from prior crashes). The CEO remarked that for the first time, he had *visibility* – a dashboard – into the e-commerce performance he could check every week, which built a lot more confidence in investing further. This is what I love to see: a company recognizing the growing importance of e-commerce, **investing in the team and structure**, and reaping the rewards.

And remember our earlier story of Rural King – the big omnichannel retailer that attempted to "go big or go home" by launching a new platform, BOPIS, PIM, and more all at once ? The launch was tough (to say the least), but it succeeded by brute force. The big lesson the Rural King team learned was to phase things next time – and to ensure the *organization* was ready. They have since embraced a more iterative approach: smaller, frequent updates, training store staff gradually, and giving their e-com team a stronger voice in planning. In other words, they moved from an ad-hoc project mindset to a steady, pod-based *product* mindset. If a giant farm and home retailer can do it, so can you.

Moving Fast: Embrace Change and Ship Results

We've covered a lot: from horror stories of siloed teams, to the nuts-and-bolts of RACI charts and pods. Now I want to leave you with a challenge and some encouragement. The challenge is to **confront your own organization's inertia**. It's easy to nod along and agree in theory, but change is hard when you're back at your desk facing quarterly targets and fire drills. You might be thinking, "This sounds great, but I've got this one VP or that legacy system that will fight me." You're right – there will be friction. But as we've said, **change is coming one way or another** in commerce. The question is whether you lead it or get dragged along. I urge you to lead.

Start with a small step. Maybe that's booking a meeting with your CEO or COO to discuss elevating the e-commerce function. Maybe it's assigning someone tomorrow as the "Product Owner" even if it's just a title change initially – and giving them a mandate to prioritize the backlog. Or it's forwarding this chapter to your team and saying, "Hey, this is what I want us to try." You don't need permission to fix what's broken. I've built my career on a bit of a rebel streak – respectfully but firmly challenging the status quo. I'm giving you permission, if you need it, to do the same. **Challenge outdated mental models** in your company. If someone says "We've always done it this way," ask "But is it working *now*?" Use data and customer feedback as your sword. It's hard for an executive to argue against facts like, "Our online conversion rate is 1%, industry benchmark is 2%, that's a $XM opportunity we're missing. I believe our team structure is part of the problem and here's how we can fix it."

And here's the encouragement: once you start making these changes, you'll likely see *quick wins*. I've had clients report that after redefining roles and empowering their teams, they delivered their **first ever on-time website launch** (after a history of delays). Another saw their deployment frequency go from monthly to weekly, which meant **more experiments and faster learning**. When you ship more often, you learn more often – and that leads to better results. Perhaps most importantly, **team morale goes up**. People love being able to do their jobs without jumping through hoops. Developers love actually seeing their code go live and improve metrics rather than dying in endless QA. Marketers love seeing campaigns launched on time. Customer service loves being heard and seeing their feedback lead to site improvements. The whole company can feel the momentum when the e-commerce engine revs up. It's exciting – this is why we call this book a *Growth Playbook*. It's about unlocking that momentum.

As a mid-market leader, you might not have the war chest or the brand clout of an Amazon or Walmart, but you have the ability to **be nimble and innovative**. Org design might not sound as sexy as the latest marketing hack or AI tool, but trust me, it's a force multiplier. It creates the environment where all those other wins can happen. It's the platform for growth.

Before we wrap, I want to arm you with a few resources. **At the end of this book, you'll find downloadable tools** to help you implement what we discussed. There's a **one-page RACI template** ready for you to fill in with your team's names, an **org chart blueprint** for the three-layer team structure, and a sample **e-commerce Scorecard dashboard** (inspired by EOS®) you can use to start tracking your weekly wins. We've also included a **Reorg Sprint Checklist** – a step-by-step list to guide you through that 8-week restructuring process I outlined. Take these resources, tweak them to fit your reality, and put them to use. They're designed as quick-start aids so you don't have to start from a blank page.

E-commerce 3.0 is here. It's fast, it's customer-centric, and it rewards the bold. By reimagining your org chart and culture around the principles we've discussed – small cross-functional teams, clear ownership, data-driven decision making, and ruthless elimination of silos – you position your business to not just survive, but thrive in this era. I'm excited for you to take this playbook and run with it. The companies that embrace these changes will be the ones shipping new features weekly, delighting customers, and hitting that next level of growth.

So go ahead – **build your e-commerce dream team**, and go seize that growth. As always, I'm rooting for you, and I can't wait to hear about the results you achieve. Now, onto the next chapter, where we'll continue building your e-commerce growth toolkit. Onward!

4.2: CHANGE MANAGEMENT WHEN EVERYONE'S TIRED

An all-too-common scenario: A once-thriving $80 million outdoor brand kicks off a massive replatforming project, confident the new tech stack will ignite growth. Six months in, 40% of the team is gone. Burnout swept through developers, project managers, even a VP. Why? The project marched forward with **no quick wins** to show for countless late nights. The remaining team members are exhausted and demoralized, questioning if the end is even in sight.

This is the **human cost** of change done wrong. I've seen it firsthand. In one case, a client's e-commerce team kept chasing the "next big platform" every year or two, convinced a shiny new system would fix everything. They spent time and money on replatforming – only to end up with a nicer-looking site and *the same old problems*. The strategy itself wasn't necessarily wrong, but even the right strategy fails if the humans carrying it run out of gas. People are the engine of execution, and when that engine overheats, *game over*. You could implement the best e-commerce platform in the world and still have the same problems if you haven't addressed your team's fatigue, processes, and morale. In other words, no change initiative succeeds if your people are too drained to drive it.

Why Mid-Market Teams Hit the Wall

Why do so many mid-market e-commerce teams hit a wall during major change efforts? In my experience leading transformations at Creatuity and advising mid-sized brands, a few culprits crop up repeatedly:

Talent Shortages: Mid-market firms often operate with **lean teams**. There's rarely bench depth; everyone is wearing multiple hats. When a big project kicks off, these teams don't have spare capacity. If one key developer or digital marketer quits (or even takes PTO), suddenly 10 other tasks stall. Hiring quickly isn't easy either – you're competing with big tech salaries and hot startups. The result is *the same few people trying to do everything,* a perfect recipe for burnout.

Remote Fatigue: A few years into the pandemic-powered remote-work revolution, many teams feel the fatigue. Yes, remote work boosted flexibility, but it also introduced isolation and communication gaps. The casual office camaraderie that could energize people is harder to capture on Zoom or Slack. Miscommunications fester when you can't just pop into someone's office. I've seen trust and cohesion drop in teams that went fully remote without strong cultural rituals. As I discussed in a podcast episode on team dysfunction, building trust is even harder in our remote world these days. When trust is low, everything slows down and people disengage. The enthusiasm to push through challenges just isn't there. Don't get me wrong - forcing everyone back into the office isn't a solution, but you have to lead your team differently when they are remote or distributed.

Change Overload: Mid-market companies have faced **non-stop change** in recent years – from pandemic disruptions to supply chain chaos to shifting consumer behavior. After pivoting again and again, teams are *change weary.* Every new platform, org restructure, or "urgent initiative" starts to blend into a blur of perpetual upheaval. I've heard it from clients: *"We're tired of 'new' – can we catch our breath?"* When people feel like the goalposts move every quarter, they eventually hit a psychological wall. They start to freeze or even quietly rebel against yet another big change. In contrast, the few that thrive in chaos find ways to balance long-term strategy with short-term wins and recovery time.

In short, mid-market teams hit the wall because they're **under-resourced, over-stretched, and running on fumes**. Understanding these factors is the first step. Now, how do we manage change in a way that doesn't break our teams?

The "Three-Battery" Model of Change Capacity

Think of your team's capacity for change as a set of three batteries: **Cognitive, Emotional**, and **Physical**. At the start of a project, all three batteries need to be charged for each team member. If any one of them hits zero, *the whole project stalls*. Here's what I mean:

Cognitive battery – This is mental bandwidth and focus. Big change initiatives demand a lot of brainpower: learning new systems, solving novel problems, creative brainstorming. If people are constantly context-switching or drowning in information overload, their cognitive battery drains. Ever see a developer stare blankly at code at 2 AM, or a marketing lead unable to make one more decision? That's a dead cognitive battery.

Emotional battery – This is morale and motivation. Change is an emotional rollercoaster. There are moments of excitement and moments of fear, doubt, frustration. Leaders often underestimate how **emotional burnout** can derail a project. If the team loses heart – say, because they don't see progress or don't feel appreciated – their emotional energy tank hits empty. They become disengaged passengers rather than drivers of the change.

Physical battery – The basic energy from rest, health, and stamina. Crunch mode culture can drain this fast. If people are pulling long hours, sacrificing sleep or exercise, their physical battery depletes. You'll notice more mistakes, bugs, and sick days. I've had a client whose IT lead landed in the hospital from pneumonia right before Black Friday – his body gave out after months of 70-hour weeks. When physical energy is gone, no amount of willpower can compensate.

A sustainable change program **monitors and recharges all three batteries**. As a leader, you need to watch for signs of cognitive over-load (confusion, analysis paralysis), emotional exhaustion (apathy, irri-tability), and physical fatigue (health issues, absenteeism). Importantly, encourage recharging: whether it's giving the team a long weekend off after a big push, bringing in an extra contractor to reduce cognitive load, or publicly celebrating small wins to refill the emotional tank. We'll talk soon about a 90-day plan to "recharge & ship," which explicitly builds in recovery. The key insight of the three-battery model is this: **if any one battery dies, the whole engine of change can shut down**. Your job is to ensure none of them hit zero.

Five Blindspots That Kill Projects

Most failing projects don't die because of bad tech or bad intent – they die by *blindspot*. After years of consulting and post-mortems, I've noticed five common blindspots that can silently wreck an e-commerce transformation. Let's shine a light on each:

"Executive teams, the cause of and solution to all of your replatforming problems." – a wise (and somewhat cheeky) truth I shared on the podcast. Leadership blindspots at the start can set the stage for failure later.

Vision Vacuum – This blindspot is a **lack of a clear "Why"**. The project exists, but no one can articulate the true purpose or urgent business outcome driving it. Without a compelling vision, teams wander. I call it *"strategy by placeholder"* – you assume there's a strategy, but really there's a vacuum. I've seen executive teams decide "we must upgrade our platform" yet fail to explain *why now*. As a result, every department starts tacking on pet features and conflicting priorities. One client's exec came back from a trade show with **15 new "must-have" ideas** for the e-commerce site – none tied to a real customer need. That's a vision vacuum at work. People will chase shiny objects if you haven't lit a North Star. Even a Ferrari goes nowhere without a destination. In practice, fill the vision vacuum by defining a single, clear **North Star goal** (e.g. "reduce checkout drop-off by 20%" or "launch in a new market by Q4"). Communicate it constantly. Every project meeting should reiterate *why* you're doing this. Know your why, and make sure it's a true urgency, a real business outcome. If the vision is clear and shared, you energize the team instead of exhausting them.

Metrics Mayhem – Another blindspot: either **drowning in too many metrics or using the wrong ones**. This is when teams can't tell if they're winning because the scorecard is a hot mess. One project had 50 KPIs on a dashboard – nobody knew which mattered, so everyone ignored it. In another case, a team only tracked vanity metrics (like marketing emails sent) but not core ones (like conversion rate or fulfillment error rate). They declared victory on "emails up" while revenue quietly tanked – oops. Metrics mayhem creates confusion and erodes trust; team members start feeling like all their work disappears into a black hole of meaningless numbers. Instead of "what gets measured gets managed", these teams seemed to live by what gets measured gets manipulated – or worse, ignored. To fix this, choose a **handful of meaningful metrics** tied directly to the project's success (e.g. deployment frequency, customer NPS, page load time improvement, etc.). Make them visible to everyone and discuss them in context. And remember, metrics should guide decisions, not serve as wallpaper. If a metric isn't helping you course-correct or celebrate progress, it's noise.

Hero Culture – This blindspot lurks in many passionate teams: the "**hero mentality**" that glorifies individuals burning the midnight oil to save the day. On the surface, it sounds positive – dedication! But hero culture is a double-edged sword. It often masks underlying process problems and **leads to burnout**. I've walked into companies where a few star players were pulling insane hours to hit deadlines, while others became bystanders. The heroes were lauded, but guess what? A year later those heroes either got poached by competitors or collapsed from exhaustion. The project timeline? In shambles because no one else fully knew what the hero knew. *Burnout is not a badge of honor; it's a project failing in slow motion.* To combat hero culture, build **redundancy and teamwork**. Ensure knowledge is shared (pair programming, documentation, cross-training) so no task relies on one savior. Reward *team wins* publicly, not individual heroics. And if you see someone constantly acting as hero-firefighter, dig into why the fires keep happening in the first place. Often it's a signal to improve planning or resourcing.

Scope Creep – Ah, the classic: **trying to do everything, ending up doing nothing well**. Scope creep is a silent killer, especially when there's no single owner reining it in. One tell-tale sign is what I call the *"wishlist committee."* Without strong scope discipline, every department and stakeholder starts adding their two cents. Before you know it, your "redesign the homepage" project has morphed into "replatform the entire ERP, launch 5 new features, and rebrand the company, all at once." I've seen this happen – and the project imploded under its own weight. Every department was adding to the wish list and they didn't have one clear person focusing on cost, scope and value. The blindspot is assuming more = better. In reality, **more scope = more strain** on your already tired team. If everything is a priority, nothing is. The antidote: define a **"minimum viable project"** – the smallest scope that delivers a meaningful result. Or as another client liked to call it, minimum delightful project - the smallest scope that would delight users and stakeholders. Assign a *scope guardian* (ideally an executive with P&L responsibility) to say **"No"** to additions that don't serve the core goal. Time-box scope decisions in 30-day intervals: at each checkpoint, consciously decide if new ideas are urgent or can wait for a future phase. By keeping scope tight, you preserve team energy to actually deliver quick wins, rather than chase an ever-moving target.

Tool Sprawl – Finally, the blindspot of **tool overload**. Mid-market companies often lack a unified IT strategy, so they accumulate SaaS tools and systems like barnacles on a ship. Marketing has its favorite CRM, ops introduced a new order management system, devs are experimenting with three different JavaScript frameworks – and none of these talk to each other. Tool sprawl creates fatigue in two ways: *training fatigue* (people constantly having to learn new interfaces, new logins, new workflows) and *integration fatigue* (data and processes fractured across systems). I consulted with a merchant that had **five separate tools** just for internal communication and ticketing – employees spent more time updating status in tools than doing actual work. Tool sprawl is a productivity tax and a morale killer, as teams feel like they're fighting the tools instead of benefiting from them. The easiest way to overwhelm a team is to 'simplify' their work by adding yet another tool. To mitigate this, audit your tech stack. Consolidate redundancies (do you really need Asana *and* Jira *and* Basecamp?). Choose tools that integrate well, and designate **one source of truth** for each domain (e.g. one project management board everyone uses). Also, pace the introduction of new tools – align them with training and clear value, so the team doesn't see it as just one more thing thrown onto their plate.

These five blindspots – Vision Vacuum, Metrics Mayhem, Hero Culture, Scope Creep, Tool Sprawl – can sneak up on any organization. The good news is that once you *see* them, you can address them. In fact, in the next section, we'll talk about a proactive model to drive change that inherently avoids many of these pitfalls.

THE CHANGE FLYWHEEL

To prevent the wheels from falling off your transformation, I advocate a six step **"change flywheel"** – a repeatable cycle that keeps projects humming and humans energized. It's a distillation of what's worked for us and our clients, and borrowing from agile practices and EOS* (Entrepreneurial Operating System) cadences. Here are the six steps, in order:

Story – It all starts with a compelling story. This is your Vision step, but it's more than a dry goal – it's the narrative that gives the project meaning. Frame the change as an inspiring story your team can get behind. For example, instead of "We're upgrading our e-commerce platform," try "We're reimagining the customer experience so it's as legendary online as it is in our store – no more lost orders, no more 5-second page loads." Share a customer anecdote or a mission-driven angle. Humans are wired for stories, not PowerPoints. When your team can *tell* the story of the change in their own words, you know you've ignited that first spark.

Scorecard – Now translate the story into **concrete success measures**. This is akin to the EOS scorecard: a handful of weekly metrics everyone can track. Define what winning looks like in numbers *and* make it visible. For example, if the story is about improved customer experience, your scorecard might include "average delivery time" or "CSAT (customer satisfaction) score" or "conversion rate". The key is to choose metrics that *directly link* to the story's promise. Keep it to 3-5 metrics so it's actionable – the team should be able to recite them by memory. Then update progress regularly (say, every Monday stand-up or in a shared dashboard). A focused scorecard keeps the team honest and aligned, avoiding that Metrics Mayhem blindspot. When everyone sees a metric move in the right direction, it's a jolt of energy; if one lags, it's an early warning system to discuss in your next meeting.

Sprint – With story and metrics in place, break the work into **short, manageable sprints**. I recommend 2-week sprints (a timeframe long enough to produce something, short enough to course-correct). Plan just enough for the next sprint – what *specific outcomes* or deliverables will we complete? Sprints create a rhythm that combats change overload: the team only needs to focus on the next two weeks, not the year-long mountain. It's psychologically freeing. Make sure each sprint includes a realistic workload (taking into account those three batteries of capacity) and builds toward a visible output. Perhaps Sprint 1 ends with a prototype, Sprint 2 ends with a pilot feature live to 5% of users, etc. Sprints also allow for built-in checkpoints to reassess scope or approach without derailing the whole project.

Shield – While the team is sprinting, leadership's job is to **shield them from distractions**. This means protecting the team from scope creep, conflicting priorities, and external noise during the sprint. In practice, it might be an executive sponsor saying "No" to an unrelated task that someone tries to slip in mid-sprint, or a project manager filtering communications so engineers aren't bombarded by every stakeholder question. I use the term *shield* deliberately – picture a phalanx guarding its soldiers as they advance. Shields up: postpone new feature requests to the backlog for later review, fend off any "drive-by" leadership mandates that aren't truly urgent, and maintain a stable environment for the sprint's duration. A practical tip: establish "office hours" or a scheduled review point for outside ideas, so people know *when* they can bring things up – and make it clear that mid-sprint is heads-down time. By shielding your team, you prevent the burnout that comes from constant context switching and you nurture that state of flow where real progress happens.

Showcase – At the end of each sprint (or at least every few sprints), run a **showcase**. This is a brief, celebratory demo or review of what the team accomplished. It serves two crucial purposes: **recognition and feedback**. Recognition means you publicly acknowledge the team's hard work and achievements – even if it's a small win, highlight it. ("Hey everyone, look what our ecomm team built in the last 2 weeks – a new feature that cut checkout time by 30%!") This boosts emotional batteries by giving a sense of progress and pride. Feedback means you invite stakeholders or a pilot customer group to experience the change and react. Keeping them involved ensures you're still on track with the Story (vision) and helps catch any issues early. Showcases create a cadence of quick wins, counteracting the demoralizing feeling of "all that work and nothing to show for it." In the earlier anecdote of the failing replatform, had they showcased incremental improvements – e.g. a new mobile UI or faster page load – maybe the team wouldn't have felt like they were on a death march with no end. Make show-cases informal, fun, and regular. Share before-and-after screenshots, let a team member narrate the journey, and maybe include a customer quote if available. This step re-energizes everyone for the next sprint.

Sustain – The final step is about **making change stick** and sustaining momentum. Often after a big release or the first major deliv-erable, teams collapse or move on without solidifying the gains. To avoid that, have a sustain plan: continue measuring the scorecard metrics post-launch to ensure the change is delivering the expected results (and not causing new issues). Set up ownership for ongoing maintenance of whatever was built – who is tending this garden so it doesn't become a jungle again? Culturally, sustain is about reinforcing behaviors: if the project introduced a new process (say, code review practice or a new customer service protocol), make sure it's docu-mented, and integrate it into training for new hires, etc. **Sustain is also about continuous improvement** – take lessons learned from this cycle and feed them into the next Story. This is why we call it a *flywheel* – once you sustain the change, you loop back to craft the next chapter of the story, maybe tackling a bigger challenge now that you have a win under your belt. One tool that helps here is the EOS concept of quar-terly "rocks" and agile retrospectives: every 90 days, reflect on what's working, what's not, and adjust. Don't mandate that everything is perfect; instead, create a culture where the team feels ownership to maintain and improve their work over time. Sustain also means keeping those three batteries charged for the next go-around – cele-brate with a team outing or some comp time if possible, to recharge before you spin the flywheel again.

By cycling through **Story** → **Scorecard** → **Sprint** → **Shield** → **Showcase** → **Sustain**, you create a self-reinforcing momentum for change. It's structured enough to prevent chaos, but agile enough to adapt. We derived this from battle-tested EOS rhythms and agile principles, and it strikes a balance: you're always aligning to the vision (story), tracking progress (scorecard), executing in focused bursts (sprint) with protection (shield), celebrating results (showcase), and institutionalizing successes (sustain). Compare that to the old way – big plan, bigger scope, endless grind, big flop – and it's clear why a flywheel beats a burnout cycle.

CULTURE HACKS THAT ACTUALLY WORK

Sometimes small cultural hacks can make a big difference in keeping your team energized through change. Not gimmicks, but *practical tweaks* that improve collaboration, learning, and morale. Here are a few I've seen actually work in mid-market e-commerce teams (including my own at Creatuity):

Two-Pizza Teams: Amazon popularized this idea – any team should be small enough to be fed by two pizzas. In practice, this means breaking big groups into **agile, autonomous pods**. For your change initiative, consider forming a dedicated strike team of 5-7 people (cross-functional) focused solely on that project. We did this with a client's replatform: rather than 20 people juggling it part-time, we carved out a pizza-team that owned it end-to-end. The intimacy and focus of a small team skyrockets accountability and communication. People feel more connected and less lost in the crowd. And if you have multiple such teams, healthy competition can spur them on. Just remember to align them with a common story and scorecard to avoid siloing.

Public OKRs: OKRs (Objectives and Key Results) are a goal-setting framework, and making them **public** (internally) can be helpful. When everyone, from the CEO to an intern, can see each team's OKRs and quarterly progress, it creates transparency and shared purpose. I advise mid-market leaders to publish the project's goals and progress on a visible dashboard or wiki. For example, if your e-commerce revamp has an OKR of "Improve mobile conversion from 2% to 3% by Q3," put that up on a big screen in the office or a Slack channel. Update it regularly. This visibility does two things: it keeps the team **focused on outcomes** (reducing random task thrash), and it fosters a bit of social pressure – nobody wants to be the one lagging on a public goal, but also everyone can rally to help if they see a target slipping. Public OKRs turn goals into a team sport rather than a private hope.

Dev-for-a-Day (Role Swap): One cultural experiment I love is a **role-swapping initiative**. For one day (or even just a few hours) a month, have team members swap roles or at least shadow another role. For example, a developer spends a day with customer service listening to customer calls, or a marketer sits with DevOps to see how deployments happen. We informally call it "Dev-for-a-Day" or "[OtherRole]-for-a-Day" depending on who's swapping. The effect is powerful: it breaks silos, builds empathy, and often sparks ideas to improve processes. One of our retail clients had the e-commerce manager answer support tickets for a day – she was shocked by some of the issues customers faced, and immediately prioritized a fix that her team had overlooked. Similarly, letting a junior developer shadow a product manager for a day gave them context on *why* certain features were requested, which improved their coding decisions. It's like a mini job rotation that rejuvenates perspective. Plus, it can be fun and breaks the monotony. Just be sure to have willing participants and frame it as learning, not an evaluation.

Microlearning Fridays: Continuous learning keeps a team fresh. But lengthy training sessions or big conferences are hard to cram in during a major project. Enter **Microlearning Fridays**. Dedicate a small block of time on Fridays (say 30 to 90 minutes in the afternoon) for bite-sized learning and sharing. It could be as simple as one team member presenting a 10-minute demo of a new tool or a cool e-commerce trend they researched, followed by casual discussion. Or use the time for everyone to take a short online course module (there are great quick courses on UX tweaks, Google Analytics tips, etc.). The key is *regular, small doses* of learning that feel like a break from routine. This practice keeps curiosity alive and gives that little end-of-week boost. People finish the week feeling like they grew their skills rather than just fought fires. And crucially, it reinforces a culture that values improvement and adaptation, which is exactly what you need during transformation.

These culture hacks – small teams, transparent goals, role-swaps, microlearning – all share a theme: **they reconnect people with purpose and each other**. They're antidotes to the fatigue that sets in when all you see is a grind. By keeping teams small and focused, everyone sees their impact. By sharing goals and knowledge, you remind folks they're in it together and that the work matters. Try one or two of these hacks and observe the energy lift in the room (even if the "room" is a Zoom grid). Culture isn't changed by posters on the wall; it changes through little habits that accumulate into new norms.

ARCHETYPE PLAYBOOKS

Not all e-commerce organizations are the same. A change management playbook should flex based on your business model. I've noticed patterns in three common **mid-market e-commerce archetypes** I work with: "Side-Car B2B," "Omnichannel," and "Pure-Play." Each has unique challenges and tactics that work best. Here's a quick comparison:

Side-Car B2B

(Traditional B2B company adding eCommerce as a side channel)

– Internal resistance from sales teams ("channel conflict" fears)

– Legacy processes not built for digital speed

– Customers used to high-touch, offline ordering

Bridge old and new: Set a *common vision* that eCom **augments** sales, not replaces it. Secure an executive champion from the B2B side. Run a **pilot eCom project** with one region or product line to get quick wins (e.g. online ordering for small accounts) to prove value. Provide extra training to traditional teams (e.g. digital skills workshops). Use **EOS tools** like the People Analyzer to ensure you have the *right people in the right seats* for digital efforts.

Omnichannel

(Retailer with both brick-and-mortar and online channels)

– Siloed teams (store ops vs. eCom) and potential turf wars

– Complex customer journeys (buy online, return in store, etc.)

– Data fragmentation between systems (POS, eCom platform, etc.)

Unify the experience: Create cross-functional **"two-pizza" teams** mixing store and eCom staff for projects (e.g. curbside pickup launch team). Align everyone on shared customer metrics (like overall NPS or total LTV) to break silos. Implement regular **joint planning meetings** –

eCom, marketing, store ops all plan campaigns together. Quick win idea: a *store associate for a day* program where HQ eCom staff spend a day on the floor to see pain points. Tech focus on integration – prioritize connecting systems (invest in middleware or an integration platform) before flashy new features. Communicate every improvement as a win for *both* online and store sales.

Pure-Play

(Digital-native DTC brand or online-only retailer)

– "Move fast" culture can lack process, risking quality issues

– Smaller team, every outage or misstep hurts more

– Tendency to chase shiny new tools (since no legacy systems holding back)

Discipline and sustainability: Introduce just enough **process** to avoid chaos (e.g. implement code freeze windows during peak season, add QA checklists) without stifling innovation. Use **OKRs** to focus the team's many ideas into a few strategic bets each quarter. Watch for burnout – in a passionate startup-like environment, the emotional battery drains quickly. So enforce downtime: e.g. a rule that everyone takes a full week off every X months, no optional overtime. Leverage the team's tech enthusiasm by running **innovation sprints** (hack weeks) to get the exploration bug out of their system, but then channel that into core roadmap execution. Essentially, tame the chaos dragon *just enough* so you can scale reliably.

Each archetype benefits from the general principles we've discussed, but the emphasis differs. For a Side-Car B2B, it's about **internal alignment and proving e-commerce's worth** to skeptics. For Omnichannel, it's breaking silos and delivering a seamless customer experience across channels (which means uniting people and data). For Pure-Plays, it's adding a layer of maturity – process and focus – to their agility so they don't flame out. Identify your archetype (or mix of them), and adjust the playbook accordingly. One size does not fit all in change management, but these patterns give you a head start.

METRICS DASHBOARD FOR CHANGE

Earlier we touched on avoiding "Metrics Mayhem." Now let's

outline a **focused dashboard** of metrics that actually gauge your team's change health. Think of these as the vital signs for your transformation project:

Change Velocity

(Features or improvements delivered per sprint/quarter)

Measures the **pace of execution**. Are we delivering value increments quickly? This combines speed and throughput of the team.

If velocity drops >20% over two consecutive sprints, or falls below historical baseline, it's a red flag. It could indicate overload, blockers, or waning motivation. Don't just push harder – investigate why.

Employee NPS

("Would you recommend working here?" survey)

Gauges **team morale and engagement**. High eNPS means people are energized; low means they're frustrated or burnt out.

eNPS turning negative or dropping by 10+ points since last quarter is a big warning. If people wouldn't recommend the team, they might be halfway out the door. Act fast – pulse survey for qualitative feedback, address common pain points (workload, recognition, etc.).

Defect Escape Rate

(% of bugs/issues discovered by users after release vs. before release)

Proxy for **quality under change**. Some increase in bugs under rapid change is normal, but too high means you're trading quality for speed.

If >15% of total defects are found by customers (or in production) rather than in testing, slow down. High escape rate signals corners are being cut or testing is insufficient. It's a morale issue too – frequent firefighting exhausts teams. Consider adding a hardening sprint or extra QA resources if this spikes.

Scope Creep Delta

(% increase in project scope or requirements over original plan)

Monitors **project discipline**. Some scope change is expected, but uncontrolled creep is measurable by how much the to-do list grows.

If scope has expanded >25% without timeline/budget adjustments, red alert. The project may be veering off the agreed path (remember the wishlist committee problem). It's time to pause and re-prioritize. Either cut back to core or officially re-baseline the plan with more resources, otherwise the team will drown.

. . .

Keep this dashboard simple – these four metrics cover execution speed, team health, quality, and scope control. Review them in your weekly or bi-weekly leadership sync. The point isn't to punish the team with numbers; it's to **catch red flags early** and spark constructive problem-solving. For example, if change velocity is dipping, ask why: is it external dependencies? Over-engineering? Too many priorities? In one client case, we saw velocity crater and discovered the team was dragged into unrelated maintenance work – a quick re-shielding fixed it. If employee NPS is dropping, have an honest talk with the team (possibly anonymous) to learn what's draining them – it could be something fixable like meeting overload or unclear direction.

One more metric I sometimes include is **"Change Adoption Rate"** – e.g. what percent of the team is using the new tool or process introduced. If you launch a new workflow but only 30% follow it, that's telling. But you can decide what's relevant depending on your project (training completion rates, customer adoption of new features, etc., could be variants).

The qualitative side matters too. Pair the dashboard with open-ended feedback. But having these key metrics gives you a balanced scorecard: speed, people, quality, scope. If all four are green, you're likely in that sweet spot where change is happening at a healthy, sustainable clip. If one goes red, you know where to dig in before it cascades into a bigger issue.

90-Day "Recharge & Ship" Plan

When everyone's tired and a project has lost momentum, sometimes you need to hit the reset button. I often recommend a 90-day "Recharge & Ship" plan – a focused three-month period to *re-energize the team* and *deliver a tangible win*. It's essentially applying the Change Flywheel with an emphasis on rest + result. Here's a week-by-week outline of how you might execute this:

Week 1: Reset and Realign – Call a timeout on the usual grind. Convene the team (in-person or virtual) for a frank retro + kickoff meeting. Acknowledge the fatigue and frustrations openly. Revisit the *Story*: why are we here, what's the vision? You might even scrap or simplify overly complex goals. By end of Week 1, define a clear, **rallying mission** for the next 90 days (e.g. "Launch the new mobile homepage by end of Q, with zero downtime"). Keep it simple and motivating.

Week 2: Recharge Prep – This week, focus on **capacity planning and quick wins to build confidence**. Adjust workloads: redistribute tasks, bring in a contractor or shift some responsibilities if possible to lighten the load on overburdened folks. Encourage team members to take a breather – maybe a 2-day mini-vacation or lighter Friday schedule. Simultaneously, identify one or two *lingering small tasks* you can knock out quickly (fix a pesky bug, deploy a minor improvement). When the team sees something accomplished this early, it generates optimism.

Week 3: Sprint 1 Kickoff – Launch the first full sprint with the renewed team. Clearly scope Sprint 1 to be **doable** in two weeks – nothing heroic. Assign roles, clarify the "definition of done" for this sprint's outcomes. Also, introduce any **team norms** changes: e.g., "No meetings after 3pm," or "quiet focus hours each morning," etc., to give people some relief in their day. By now everyone should feel, *"Okay, we have a plan and the air cover to actually do it."*

Week 4: Sprint 1 Execution – Heads down, getting stuff done. As a leader, practice the **Shield** step: guard the team's time. Check in mid-week not with "are we on track?" pressure, but "any blockers I can remove?" encouragement. Also keep an eye on working hours – if someone's still pulling late-nighters, step in to balance the load. End of Week 4, do a quick showcase internally: what did we complete? Even if it's rough, have a demo. Give kudos in the team chat or at Friday stand-up. Small celebration (virtual high-fives, maybe e-gift cards for a coffee) to mark the completion.

Week 5: Sprint 2 Planning – Take lessons from Sprint 1. If it went great, keep the formula; if not, adjust. Perhaps Sprint 1 revealed some overestimation – so dial back Sprint 2 scope slightly. Or maybe someone discovered a new risk – address it now. Use Week 5 to plan Sprint 2 in detail. Also, ensure **recharge activities** are ongoing: maybe schedule a team lunch (even on Zoom) to socialize a bit, or share some funny memes about the project in a group chat to lighten the mood. This maintains the human connection.

Week 6: Sprint 2 Execution – By now a rhythm is forming. The team is hopefully gaining confidence from the Sprint 1 win. Keep shielding and motivating. Introduce a fun element mid-sprint: for example, "Feature Friday" where the team gets to spend half the day experimenting with a solution to a problem they care about (aligned with project). It might yield a great idea, or just give a mental break. End of Week 6, showcase again – maybe this time include a couple of additional stakeholders or a friendly customer to get external positive feedback.

Week 7: Midpoint Retro & Rest – We're about halfway (day 45). Pause for a **midpoint retrospective**. What's working? What's not? Crucially, gauge the batteries: cognitive load okay? Emotional mood? Physical hours? Adjust plans for the next sprints accordingly (maybe we need to cut a feature or add a person). After the retro, **enforce a little rest**: consider giving the team an extra day off or a low-key work-from-home Friday with no meetings. This mini-recharge keeps them fresh for the second half.

Week 8: Sprint 3 Planning – Plan the next sprint with the insights from the retro. By now you should also firm up what the "big ship" at Day 90 will be – likely a culmination of sprints 1-4 deliverables. Ensure Sprint 3 tasks are aligned to hit that target. This week, also start **planning the final testing/release** needs if your project involves a launch (begin booking time with QA, scheduling a soft launch window, etc.). Communication is key: update higher-ups on the progress so far – share those metric dashboard results if they're looking good (build confidence and buy-in so nobody derails you in the home stretch).

Week 9: Sprint 3 Execution – This is often the sprint where unexpected things pop up (integration issues, a team member out sick, etc.). Stay agile and supportive. Double-down on **Shielding** – often as you near a delivery, other departments suddenly realize "oh can you also include X?" or someone higher up in the company might try to add scope. This is the moment to hold the line (politely but firmly). Remind everyone of the mission and Day 90 goal. Keep the team focused. End of week, do the showcase again – by now it might be a near-complete feature or a measurable improvement. Hype it up! If Sprint 3 was tough, ensure people know what they achieved despite the challenges.

Week 10: Hardening & Sprint 4 Planning – With most of the heavy lifting done, Week 10 can serve as a *hardening sprint* or buffer. Final testing, bug fixes, content prep, etc., often happen here. Plan Sprint 4 to basically be **final tasks and launch prep**. It might not be a full two-week sprint if the team is ready to ship sooner – adjust timeline if you can deliver early. Importantly, start organizing the **big showcase or launch demo** that will happen at the end of Sprint 4. Invite leadership, make it a bit of an event. People love a deadline when there's a real audience.

Week 11: Sprint 4 Execution (Final Sprint) – The last push. Emotions can run high – excitement and anxiety. Keep everyone calm and on target. Host a daily stand-up this week if you weren't already, to quickly surface any last-minute issues. Be ready to cut any remaining "nice-to-haves" – better to deliver a solid core than fumble because of an extra bell/whistle. The mantra for the week is **"Done is better than perfect, as long as it works."** Also, start prepping any documentation or training if this change affects other teams or customers (e.g. if launching a new front-end, prepare FAQs for customer support).

Week 12: Launch & Celebrate – Time to **Ship**! Early in Week 12, execute the go-live or final implementation of the project. Ideally, do this early in the week to have time for any issue resolution. Once it's out there, take a moment to congratulate the team. Seriously – don't rush on to the next thing. Reflect on what was achieved in these 90 days. Hold that big **Showcase Demo** for the broader org: let the team show off the new feature/process, share before-and-after metrics if available (e.g. "page load is 2x faster now"). Encourage applause. Collect any immediate positive feedback (it's fuel for the team's emotional battery). Then, do a final retrospective purely focused on *how the 90-day recharge plan went* for the team. What did we learn about how we work best? Finally, **celebrate**: maybe it's a team happy hour or an offsite day or even a celebratory Amazon gift card for each person – something to mark the accomplishment. They've earned it, and ending on a high note is critical to sustain momentum.

By the end of this 90-day cycle, you've hopefully achieved two things: **a recharged team and a shipped result.** The team saw that it's possible to turn things around, and they have a win they can point to. That does wonders for confidence. One of my clients, a DTC apparel brand, did a plan like this – they had been stuck in analysis paralysis for months; we reset, delivered a new mobile site in a quarter. The kicker: employee satisfaction scores jumped, and the team asked, "When's our next 90-day cycle?!" Success breeds eagerness for more success, when done in a healthy way.

CASE STUDIES

To see how these principles play out in the real world, let's double-click on two anonymized case studies: one from an **outdoor gear retailer**, and one from a **DTC beauty brand**. Both were mid-market companies facing the "everyone's tired" dilemma, and both managed to turn things around.

OutdoorCo (Outdoor Gear Retailer, ~$100M revenue): OutdoorCo is a company with a proud brick-and-mortar heritage – think camping and hiking equipment, with 50+ stores and a growing online channel. They embarked on a digital transformation to unify their in-store and online experience. Halfway through, the project was in trouble. Store managers were complaining, the e-commerce team was pulling 60-hour weeks, and a wave of resignations hit the digital department (sound familiar?). The blindspots were classic: *Vision Vacuum* (store staff didn't understand why the new system was needed), *Hero Culture* (a few IT folks were firefighting everything), and *Scope Creep* (the project ballooned to include a new loyalty program and POS upgrade at the same time). When I came in to advise, the first step was **re-scope and refocus**. We pulled back to an MVP: launch *buy online, pick-up in store (BOPIS)* functionality in 3 months – a tangible win that also bridges online and offline. We formed a Two-Pizza Team with a couple of store reps and ecom folks together (breaking the silo), and set a clear 90-day pilot plan. Crucially, we got the COO (who oversaw stores) to be the **executive P&L owner** of this project, aligning it to store sales goals. Over 90 days, using the flywheel approach, the team delivered BOPIS for 10 pilot stores. The result? Online sales picked up 5% in those regions and customer satisfaction improved (people loved the convenience). Even more, the store managers – initially skeptics – became champions because it drove foot traffic. The success story spread internally, refilling the emotional batteries of the whole team. They could *see* progress and its impact. Employee turnover, which had spiked to 20% that quarter, fell back to normal. One team member told me, "I finally feel like we're winning, not just working." OutdoorCo then rolled out BOPIS nationwide with the momentum from the pilot. By focusing on a quick win, protecting the team, and fostering cross-team culture, they managed to revive a faltering transformation. Within a year, they went on to tackle inventory visibility and a mobile app – but with a much healthier team and a playbook for execution.

GlowUp Inc. (DTC Beauty Brand, ~$50M online-only): GlowUp is a fast-growing direct-to-consumer beauty brand that found success through viral marketing and influencer hype. However, internally it was chaos: a small team doing everything, flying by the seat of their pants. They launched new products at breakneck speed – which was great until a major website crash during a product drop and a fulfillment meltdown on the back end. The team was *beyond exhausted*. When every launch felt like a fire drill, their best people started leaving for saner jobs. GlowUp's CEO realized they needed more structure without losing agility. We identified blindspots: *Metrics Mayhem* (they tracked social media likes obsessively, but not supply chain KPIs), *Tool Sprawl* (10+ different SaaS tools causing errors), and of course *Hero Culture* (a couple of "rockstar" employees were the glue holding everything together, barely). The prescription was to introduce some *process and focus* to this Pure-Play (as outlined in our archetype playbook). Step 1: We implemented **public OKRs** for the next quarter – one objective being "Zero downtime on launches" with key results around site uptime and order success rate. This shifted the mindset from adrenaline-fueled scrambling to *planned reliability*. Step 2: We enforced a **code freeze** period before big launches (no last-minute feature changes 48 hours prior, something I always preach for stability). Step 3: Cross-training the team – we used the **Dev-for-a-Day** concept; the marketing folks sat with customer support during a mini launch to see the fallout of technical issues, and the engineers joined a marketing planning meeting to understand why they always asked for changes last-minute. The empathy built was huge – one developer said it was eye-opening to see how a promo tweak can lead to thousands of support tickets if the site falters. Over 3 months, GlowUp slowed down a bit to stabilize: they skipped one monthly product drop to focus on fixing known bugs and consolidating from three marketing tools down to one. The next launch, they did as a 90-day "Ship" project with full testing – and it was smooth. *No site crash, no angry Instagram rants from customers.* The team, for the first time, got to breathe easy during a launch and even take a half-day off afterward to celebrate – previously unheard of. The CEO noticed a change: "People aren't on edge anymore; they're actually excited about the next launch, not dreading it." Metrics improved too: defect escape rate fell by 40% (fewer issues reaching customers) and employee NPS jumped from negative territory to +30. GlowUp learned that a bit of discipline and a sustainable pace didn't kill their vibe – it saved their company from imploding. They now operate with a cadence of monthly mini-sprints and quarterly big pushes, and attrition has dropped considerably.

Both these cases highlight an important truth: **turnarounds are possible, even when the team is drained**. By applying the right framework (vision, quick win focus, team empowerment) and adjusting to their context (omnichannel vs. pure DTC), these companies re-energized and succeeded. Use these stories as inspiration: if OutdoorCo and GlowUp could do it, so can you. The common thread is treating the people as the most critical asset in the transformation – when you do that, the odds of success multiply.

Anti-Patterns & Quick Pivots

In the journey of change management, I'd be remiss not to warn you about a few **anti-patterns** – well-intentioned tactics that often backfire – and how to pivot away from them quickly if you see them happening.

All-Day Retrospectives: The team is tired, so leadership decides, "Let's have a mega-retrospective meeting to hash out everything." They schedule a full-day (or multi-day) retrospective hoping to solve deep issues in one go. This is usually a *mistake*. Marathon retrospectives become gripe sessions that exhaust people further and rarely produce actionable insight after hour two. **Quick Pivot:** Instead of one massive retro, break it into *bite-sized feedback loops*. Do a 1-hour retro after each sprint or milestone with focused questions (e.g. one hour just on "What's one thing that's draining us? One thing giving energy?"). You'll get more honest input and the team won't feel like they lost a day to a talk-fest. In short, keep retrospectives **short, focused, and frequent** rather than infrequent marathons.

Mandated Tools from On High: This anti-pattern is when a leader reads an article or hears about a shiny new tool and forces the team to adopt it overnight. "We're moving to X project management software next week – deal with it." The intention might be to improve productivity, but top-down tool mandates breed resentment and often **don't stick**. I recall a client whose CTO bought an expensive analytics platform to "help the team" – but never consulted the team. It sat idle for a year, money wasted, and the team felt ignored. **Quick Pivot:** If you sense a tool mandate has flopped (people aren't logging in, or complaining), *pause*. Go to the team and acknowledge, "Maybe we moved too fast on this tool. Let's evaluate together what we actually need." Pilot the tool with a small subgroup of enthusiasts to prove value, or provide proper training and support time. Even better, involve team representatives in tool selection from the start. A useful mantra: **process and tools should serve the team, not the other way around**. If a tool isn't gaining traction, it's on leadership to adapt or scrap it.

Surprise Launches: This anti-pattern is unfortunately common: in a bid to "wow" customers or executives, leadership keeps a big change under wraps until a dramatic reveal or launch. The team might be told to keep quiet or only a small inner circle knows, and suddenly *boom* – a new site or feature is rolled out without broader preparation. Surprise! The problem is, surprises are for birthdays, not deployments. Surprise launches often ignore input from those who could have spotted issues, and they leave end-users and support teams scrambling. I've seen a CEO decide to unveil a new homepage design at a board meeting without telling the customer service team – who were then flooded with confused customer inquiries. **Quick Pivot: No surprises** – pivot to *transparent, phased rollouts*. Even if you want an external "wow" moment, internally everyone who is impacted should know it's coming and be prepared. Do a soft launch or beta with a subset of users first. If you've accidentally sprung a change on the organization or customers and it's causing chaos, own the mistake. Communicate quickly: provide FAQs, training, revert if needed. Turn a surprise launch into a learn-and-correct moment. It's far better to slightly spoil the surprise than to have a debacle because people weren't ready.

"One and Done" Training: One more to note – a pattern where during a change, the company holds a single big training session (maybe a long workshop or a lengthy documentation drop) and then considers the team fully trained forever. Inevitably, people zone out or new hires miss it or details fade, and then folks struggle using the new system or process. **Quick Pivot:** Embrace **continuous learning** (like Microlearning Fridays or ongoing coaching). If you did one big training and it didn't stick, follow up with refresher sessions, short videos, or a buddy system where early adopters help others. The pivot is to see training not as a checkbox but as an ongoing part of the change rollout.

In summary, beware of well-meaning approaches that inadvertently add fatigue: *overlong meetings, forcing tools, shocking surprises, and one-off trainings*. If you catch yourself in one of these anti-patterns, don't be afraid to stop and change course. Your team will likely breathe a sigh of relief. The humility in saying "this isn't working, let's try a different way" can actually increase trust. Quick pivots, when done transparently, demonstrate that you're truly putting the team's well-being and effectiveness first, which is the whole point of change management when everyone's tired.

You've now got strategies to manage change and keep your team's batteries charged – but how do you know if all this effort is really paying off for the business? In the next chapter, we'll tackle **Metrics That Matter**. We'll dive deeper into building an executive dashboard that ties together project outcomes, team health, and business impact. It's all about measuring what truly drives growth (and avoiding vanity metrics traps). We'll also explore how to communicate these metrics to stakeholders in a way that keeps everyone aligned and invested in continued improvement.

So, as we wrap up our focus on change management and revitalizing your team, get ready to zoom out and look at the bigger picture of performance. **Up next: making sure you're tracking the right numbers (and behaviors) to sustain your e-commerce growth.** It's time to connect the human side of change with the hard data – because when you can quantify success, you can repeat it and scale it. Stay tuned for Chapter 4.3, *Metrics That Matter*, where we'll equip you with the analytical lenses to complement the leadership tools you gained here. Onward!

4.3: METRICS THAT MATTER: FROM VANITY TO VALUE

once onboarded onto a client's team to find their Slack blowing up with nearly 30 different metrics every hour. Page views, bounce rates, ROAS, NPS, CTR – you name it, it was pinging nonstop. It was like an *analytics Times Square*: bright lights and data everywhere, but no clear direction. The team was drowning in numbers. One week they'd chase a spike in bounce rate, the next they'd panic over a dip in Facebook CTR. Projects started, paused, and got canceled on a whim because every new stat triggered a debate. And in all this thrash, the **real** red flags were missed – they failed to notice their repeat purchase rate had quietly tanked 20% until it was far too late. That was the gut-punch. This client was *obsessed* with tracking everything, yet they had no true **scoreboard** to tell them if they were winning or losing. It was a costly lesson in the difference between **vanity metrics** and the metrics that actually drive value.

VANITY METRICS

Let's talk about vanity metrics – those seductive numbers that make us feel good but don't actually move the needle. I'm talking about things like raw page views, social media impressions, or a super-ficial "average ROAS" that masks unprofitable ad spend. These metrics look pretty in reports, but they're often just noise. Sure, a dashboard full of rising page views can make you feel like a marketing genius, but ask yourself: **if that number changes, do you know what action to take?** If not, it's probably vanity, not value.

I've seen e-commerce leaders burn through cash chasing the wrong numbers. One brand bragged about a 10:1 Return on Ad Spend – only to reveal they achieved it by slashing their ad budget to near-zero. Their ROAS looked fantastic, but sales flatlined because they weren't reaching new customers. That's a vanity trap: focusing on a ratio that looks "good" while the business starves. Another common sinkhole is analysis paralysis. When you're monitoring two dozen metrics, you risk a culture of reactive knee-jerk moves. A tiny dip in one metric sends the team scrambling in a new direction, often undoing good work in progress. I call this **"KPI roulette"** – every week a new metric-of-the-moment steals focus. It's exhausting, demoralizing, and worst of all, it's expensive. Time and money get wasted chasing flukes and noise instead of meaningful improvements.

The truth is you can actually be *too* data-driven – or rather, too **data-drowned**. More data isn't better if it isn't the *right* data. When your dashboard looks like a cockpit full of dials but there's no steering wheel, you're just as likely to crash as to get where you want to go. Vanity metrics tend to be lagging indicators or purely descriptive. They tell you what happened in the past, but not **why** it happened or how to improve going forward. To turn data into dollars, you need to move from vanity to **vitality**. Vital (or value) metrics are the ones that indicate the health of your growth engine. They are actionable, leading indicators that predict success and inspire specific action when they change. In short, we need to stop collecting metrics like souvenirs and start using a focused set of metrics as a **scoreboard** to drive decisions.

Metrics Pyramid: 5 Hero KPIs + Supporting Metrics

To escape the vanity metric trap, I advocate a *metrics pyramid*. At the top are your **Hero KPIs** – the five (yes, only five!) metrics that matter most for your e-commerce growth. These are the numbers that every leader in your organization should know by heart, the ones you obsess over week in and week out. Think of them as the *vital signs* of your business. Just like a doctor checks your pulse, blood pressure, and temperature to quickly gauge health, your Hero KPIs instantly signal the health of your online store.

Everything else – and I mean everything – lives in the second layer as supporting metrics. Those supporting metrics still have a purpose, usually owned by individual teams (marketing, ops, product, etc.) to troubleshoot and optimize their domains. But they are **not** on the CEO's weekly scorecard. They're more like diagnostics under the hood, whereas the Hero KPIs are the speedometer and fuel gauge on your dashboard. This layered approach keeps your leadership focus crystal clear. It also unites the team: when everyone rallies around the same handful of KPIs, you create a shared language and goal set. There's no confusion about what winning looks like – it's right there on the scoreboard.

Crucially, each Hero KPI should be **actionable**. A good gut-check is to ask, *"If this number goes up or down, do we immediately know what levers to pull?"* For example, a drop in conversion rate is actionable – you can investigate checkout issues, site speed, or pricing. But a drop in "website visits" by itself? Not so much – it begs more questions (visits from whom? from where?) and doesn't tell you what to fix. By choosing the right high-value metrics, you ensure your scoreboard isn't just wall art but a real tool to drive decisions. Fewer metrics at the top also means less room to hide from reality. When a Hero KPI goes red, it's obvious to everyone, and you can swarm on it. No more hiding behind vanity stats while the business bleeds elsewhere.

BUILDING YOUR 5-NUMBER SCORECARD

So what are the *right* five numbers? Through experience and running on the Entrepreneurial Operating System (EOS) framework, I've found that most e-commerce businesses can be run on a **five-metric scorecard**. These five "hero" metrics cover the full customer lifecycle and your store's performance. In EOS lingo, this is your weekly **Scorecard** – five numbers, updated every week, reviewed in one quick huddle. Here are the five I recommend and use with clients (with a slight twist that I'll explain):

Qualified Sessions – This measures *quality traffic*. Instead of total sessions (which can be a vanity metric and influenced by bot traffic and other nonqualified traffic), we track sessions that actually engage.

For instance, a "qualified" session might mean a visitor who spends at least 30 seconds on site or clicks beyond one page. It filters out the bounces and bots and tells you if you're attracting people who *might* buy. *Example Baseline:* say 50,000 qualified sessions last week. *Target:* 55,000 (a 10% lift via better marketing and SEO). If this number moves, we know it's time to examine our traffic sources and content – more qualified traffic means more potential sales.

Conversion Rate (Desktop & Mobile) – I cheat a little here by splitting this into two metrics: desktop conversion rate and mobile conversion rate. Today, mobile often accounts for the majority of traffic, but it typically converts lower than desktop. Tracking them separately is key – a 3% overall conversion rate could be hiding a mobile conversion problem. *Example Baseline:* Desktop 4.0%, Mobile 2.5%. *Target:* Desktop 4.5%, Mobile 3.0%. If mobile conversion lags, we might need to optimize our mobile site or simplify checkout for small screens. **(Hero hint:** If you insist on one number, track overall conversion, but keep an eye on the device breakout in supporting metrics.)

Average Order Value (AOV) – This is the average amount each customer spends per order. It's an efficiency metric: how much revenue do you generate each time you convince someone to buy? *Example Baseline:* $75 AOV. *Target:* $82 AOV next quarter. If AOV dips, we dig into our product mix, pricing, or upsell offers. A rising AOV might come from successful cross-sells ("Customers also bought…") or a new free-shipping threshold encouraging bigger baskets. It directly impacts revenue – higher AOV means more dollars from the same number of orders.

60-Day Repeat Purchase Rate – Essentially, this gauges customer retention in the near term. It asks: what percent of customers place a second order within 60 days of their first? For brands with loyal followings or consumable products, repeat rate is a huge driver of lifetime value. *Example Baseline:* 18% of new customers reorder within 60 days. *Target:* 25%. If this metric moves, it signals something about customer experience and retention efforts. A drop might tell us our post-purchase follow-up or product quality is slipping. A boost could mean a new loyalty program or email campaign is working. **Note:** Depending on your business, you might tweak the timeframe (30-day,

90-day) – the key is to capture repeat behavior relatively soon after acquisition.

Deployment Frequency – This one surprises people, but I firmly believe it belongs on the scorecard: how often are you deploying improvements to your website? In a competitive market, if you're not continuously enhancing your site, you're falling behind. We measure the number of production deployments (site releases) in the last 30 days. *Example Baseline:* 1 deployment/month. *Target:* 4 deployments/month (roughly one per week). If this number is zero or red, it means the team isn't shipping improvements – maybe they're bogged down or playing it too safe. A consistently green deployment metric indicates an **innovative**, agile culture, which is critical for long-term growth. (Of course, if you have 100 deploys a month, that might signal chaos – we're looking for a healthy cadence, not frantic patching.)

These five metrics form a robust e-commerce scoreboard. They cover the funnel from traffic (qualified sessions) → conversion → revenue per order (AOV) → retention (repeat rate) → innovation/operational agility (deploys). For many mid-market brands, this "fab five" is a great starting point. You can certainly substitute one or two to better fit your business, but resist the urge to go beyond five metrics total on the leadership scorecard. Five forces you to choose what matters most.

To make these metrics truly actionable, we add a few more EOS ingredients. **Each metric gets an owner** – a single, named human being who is accountable for that number. Not a department, not a committee, not a software tool. One person's name next to each KPI. Why? Because when everyone owns something, no one owns it. If "website conversion" belongs to the whole team, it's too easy for everyone to shrug when it slips. But if Jane is the conversion rate owner, you can bet she'll be watching that number like a hawk and rallying resources whenever it dips. Ownership creates focus and urgency. It turns metrics from abstract numbers into personal missions.

Next, for each KPI we establish **Red/Yellow/Green thresholds** – basically defining what "good" or "bad" looks like for your business. For example, we might say our mobile conversion rate is Green above 3.0%, Yellow between 2.5%–2.9%, and Red below 2.5%. These thresholds should be based on a mix of historical baseline and industry benchmarks. The idea is that at a glance, the scorecard shows color-coding: if all five are green, you're on track; if anything's yellow or red, it demands attention. And here's the rule: **if a metric turns red, we stop everything and address it.** No excuses, no "let's wait and see next week." A red metric means an emergency huddle to diagnose the issue and execute a fix or experiment to improve it. Yellow is a warning – a heads-up to start investigating before it gets worse. Green, of course, means celebrate and keep doing what you're doing. These colors create a common understanding across the leadership team. I've found it builds a culture of *determined* response: we don't ignore problems, we face them head on. It's amazing how fast you can solve an issue when it's starkly highlighted in red for all to see.

Before we move on, let me emphasize the **keep it simple** mantra. Your five KPIs and their definitions should fit on one screen with no scrolling. Imagine pulling up your scorecard dashboard in a meeting and everyone can see everything at once. That visual brevity forces discipline. We're not diving into deep analysis here – save that for one-on-one problem solving. The scorecard is a quick pulse check, not a data dump. I've seen teams try to cram standard deviations, regression curves, and 18-month trend lines for each metric onto their scorecard. It ended up looking like a NASA launch dashboard – impressive, but unintelligible and not helpful for a quick huddle. Don't do that. A simple trend sparkline or week-over-week delta next to each metric is fine, but the core is just the number and its color status. **No fluff, no filler.** Remember, the scorecard's job is to tell you if you're winning or losing, fast.

Formulas and Benchmarks: Getting to Know Your Hero Metrics

Now that we've chosen our Hero KPIs, let's clearly define them and talk benchmarks. Part of turning metrics into action is knowing exactly how they're calculated and what "good" looks like in your space. Here's a quick run-down of each hero metric, its formula, and some benchmark context:

Qualified Sessions (% of Engaged Visits): We define a *qualified session* as a visit that meets a minimal engagement threshold – for example, at least 30 seconds on site or at least one click to a second page. You can calculate a **Qualified Session Rate** = (Qualified sessions ÷ Total sessions) × 100. This tells us the quality of our traffic. Benchmark: Many e-commerce sites find that around 40-60% of their total sessions are "qualified" by such measures. (Paid traffic might be on the lower end ~40%, while organic or email traffic might be higher quality ~60%+.) If your qualified rate is much lower, you're likely spending money to bring a lot of unengaged visitors – time to tighten your targeting or improve your landing pages. The goal is to drive this percentage up, or the absolute number of qualified sessions, since those are the ones that can convert.

Conversion Rate (Desktop & Mobile): Formula is straightforward – **Conversion Rate** = (Orders ÷ Sessions) × 100 (you'd calculate this separately for desktop and mobile sessions). Industry benchmarks vary widely, but generally desktop e-commerce conversion might be in the 3-4% range, while mobile lags around 1.5-2.5% for many retail sectors. What's "healthy" is very context-dependent: a luxury goods site might only convert 0.5% of visitors and still be fine due to high AOV, whereas a fast-fashion site might need 5%+. The key is to know your historical baseline and strive to beat it. Even a 0.2% increase in conversion can translate to huge revenue gains. Use tools like Google Analytics or your ecommerce platform's analytics to compare your conv rates to industry benchmarks (just be sure you're comparing apples to apples – mobile vs mobile, etc.). We set our R/Y/G thresholds based on a combination of past performance and aspirational improvement. In our earlier example, <2.5% mobile conversion was Red because historically we hovered around 2.7%; >3.0% was Green as a stretch goal. **Tip:** If you don't know where to set thresholds, start with small increments around your baseline (e.g., ±0.2% for conversion) and adjust as you gather data.

Average Order Value (AOV): Formula: **AOV** = (Total revenue ÷ Number of orders). This metric often correlates with your product pricing – higher-price catalogs have naturally higher AOV. But whatever your starting point, the idea is to find ways to boost it: product bundles, volume discounts, free shipping thresholds, upsells, etc. AOV can fluctuate seasonally (holidays might drive it up as people splurge, or a sale might drive it down if discounts are heavy). Benchmarking AOV is tricky because it's so business-specific (a jewelry store might have $200+ AOV, a home goods store $50). Instead of industry comparables, I focus on **incremental improvement**: try to raise your AOV by say 10-15% over last year's average. If you succeed, that's extra revenue without acquiring a single new customer. On our scorecard, we might mark AOV Red if it dips more than 5% below the rolling 3-month average (indicating something's off, like maybe we discounted too steeply). Green might be hitting a new high or a specific dollar target tied to our financial plan.

60-Day Repeat Purchase Rate: This is a retention metric. To calculate, pick a cohort of new customers (for example, all first-time customers acquired in January) and see what percentage of them placed a second order within 60 days of their first. You can track it on a rolling basis (cohorts by week or month). **Repeat Rate** = (Repeat customers in period ÷ Total new customers in period) × 100. Benchmarks depend on your product type. For consumables or replenishment goods (think coffee, vitamins, pet food), you might expect a higher short-term repeat rate, say 20-30%, because customers need to re-up. For durable goods or seasonal items, 10-15% might be more typical – people aren't buying a new sofa every two months. As a rule of thumb, a 60-day repeat rate above 20% is excellent for most D2C brands unless you're specifically in a subscription-driven niche. If your repeat rate is in single digits, you have a retention problem or a product that just doesn't lend itself to frequent purchases (in which case, maybe this KPI should be something else for you, like a 1-year repeat rate or customer lifetime value). Our scorecard threshold might say Green if we're above, say, 25% (for a consumables brand), Yellow if 20-25%, Red if <20%. That signals when we need to pour energy into retention campaigns, subscription programs, or customer experience fixes.

Deployment Frequency: Calculation here is simple count of deployments (code releases to your live site) in the last 30 days, or you could do number of days between deployments as an inverse metric. Benchmarks are less about industry and more about internal goals. For many mid-sized e-commerce teams, releasing something new once a week is a great cadence. So you might set Green at 4+ deploys a month, Yellow at 2-3, Red at 0-1. I encourage teams to make even small improvements count: fixing a UI bug or adding an FAQ page absolutely *counts* as a deployment. It's not about pumping out risky changes; it's about continuous improvement. If this metric stays Red (no releases) for a couple months, it's a big warning sign – perhaps your development process is stuck, or you're under-resourced, or worst of all, you've grown complacent. On the flip side, a consistently Green deployment metric creates a mindset of always moving forward. It keeps the team *innovative* and adaptable. One caution: if you find yourself doing many emergency hot-fix deploys (to patch problems), that might inflate your count – if that's the case, consider tracking "planned feature deploys" separately from incident fixes.

One more thing on benchmarks: don't obsess over *industry average* for its own sake. Use benchmarks to sanity-check your targets and thresholds, but remember your goal is to beat **your own** baseline. As long as your KPIs are moving in the right direction (from Red to Yellow to Green) consistently, you're on the right track. Metrics are your **GPS**, not your destination. They tell you where you are and if you're headed north or south, but *you* set the destination based on your business goals.

LINKING THE SCOREBOARD TO BUDGET PLANNING

A beautiful scorecard with clear KPIs is useless if you don't put resources behind it. This is where leadership needs to tie metrics directly to money. In fact, when done right, your budget planning should be an extension of your scorecard – every dollar you plan to spend on growth should connect to moving one of those five metrics. In practical terms: if one of our Hero KPIs is lagging, we allocate budget to improve it; if it's on track, we allocate budget to scale it further.

A good rule of thumb I use is the **10–15% Gross Profit Reinvestment** model. Analyze your projected gross profit for the year and earmark about ten to fifteen percent of it for e-commerce improvement initiatives (excluding your regular marketing ad spend). This isn't a random number – it comes from benchmarking industry leaders. Publicly traded e-commerce companies typically reinvest around 10%

of their gross profit into their e-commerce platforms and technology, and smaller high-growth retailers often invest closer to 15%. That means if you expect $5M in gross profit, you'd set aside $500k–$750k for projects, optimizations, and experiments aimed at boosting your core KPIs. Think of this as your "growth war chest."

How do we apply that war chest? **Directly in service of the KPIs.** For each of your five Hero metrics, brainstorm initiatives that could improve it, then budget for the ones with the highest impact. For example, if conversion rate is yellow or red, you might fund a site speed optimization project, a checkout UX revamp, or hire a CRO (Conversion Rate Optimization) specialist – those dollars go explicitly toward lifting conversion. If AOV is lagging, perhaps budget for a bundling feature or a loyalty program that rewards bigger baskets. If repeat purchase rate is low, allocate money to a subscription program launch, a better email remarketing tool, or enhanced customer service follow-up. **Every line item in your e-commerce budget should have a KPI attached to it.** This ensures you're not just spending for the sake of it – you're investing with purpose.

Let me give you a leadership perspective: when you tie budget to KPIs, you also get clearer buy-in from finance folks (even the CFO). Instead of saying "We need $200k for a website redesign because it's outdated," you say "We're investing $200k to raise conversion from 2.5% to 3.0%, which will drive an extra $X in revenue – here's the math." Now your budget is telling a growth story. It's hard for a CEO or CFO to object to funding something when you draw a straight line from dollars to data to results. This narrative budgeting is powerful. I've seen teams that once struggled to justify expenses suddenly get fast-tracked approvals because they reframed requests around KPI impact.

Also consider **scenario-based budgeting** around your KPIs. Plan a base scenario, but also a lean scenario and an aggressive scenario. For instance, "If conversion stalls or economy softens, our backup plan is to double down on retention (maybe shifting budget to loyalty perks to protect repeat purchase rate). If we exceed our revenue plan, our aggressive plan is to pour extra into acquisition to get even more quali-fied sessions while maintaining our ROAS thresholds." By aligning these scenarios with KPI triggers (e.g., "if traffic volume falls below X, then activate Plan B marketing spend"), you ensure you're ready to act no matter what. This flexibility, grounded in your scoreboard, keeps you *united* with your finance team on when and why to spend.

Finally, keep tracking the ROI of these investments. If you put $50k into a new upsell feature and see AOV rise by 10%, calculate what that means in gross profit and report it. Over time, you'll build a case that investing that 10-15% of gross profit into the right metrics yields a healthy return, creating a virtuous cycle: improved metrics → more profit → more fuel to invest again. That's how you turn a scoreboard into a growth engine.

THREE KPI TRACKING ANTI-PATTERNS TO AVOID

Before we wrap up our metrics game plan, I want to highlight a few **anti-patterns** – common mistakes or bad habits in metric tracking that can undermine your efforts. I've touched on some of these, but they're worth calling out explicitly. Steer clear of these, and you'll save yourself a lot of pain:

Lagging-Only Focus: This is when a team tracks only big lagging indicators like total sales, total profit, or monthly revenue – and nothing else. These are outcome metrics that tell you results after the fact, but they arrive too late to inform daily or weekly decisions. If all you track is monthly revenue, by the time you realize you missed your number, the month is over and you can't course-correct. It's like driving by looking in the rear-view mirror. **Avoid the lagging-only trap by ensuring your scorecard has leading indicators** (like conversion rate, traffic quality, etc. that predict revenue). You still watch the lagging outcomes, but they're not the only gauges on your dashboard. The Hero KPIs we picked are largely leading or real-time indicators, giving you time to react and improve the outcome before it's set in stone.

"KPI Roulette" (Constantly Changing Metrics): I've seen leadership teams that treat metrics like flavors of the week. One week the VP is all about reducing bounce rate, the next month it's "we need to focus on customer lifetime value," then a board member asks about NPS and suddenly that's the new north star. Changing your KPIs too often is a surefire way to confuse and frustrate your team. It takes time for a metric improvement effort to bear fruit. If you swap your Hero metrics every quarter (or worse, every meeting), you'll spin in circles and never see sustained improvement. **Pick your five and stick with them for at least 90 days**, if not a full year. Minor tweaks are fine (maybe you learn a different metric definition is better), but don't play roulette by constantly redefining success. Consistency builds mastery – your team will get better at moving a metric the longer they live with it. Also, by holding steady, you get to actually see trends and patterns. KPI roulette, on the other hand, is just chasing shiny objects and fad metrics, usually to please whoever shouted loudest last.

Metric Gaming and Vanity Tactics: The final anti-pattern is gaming the metrics – i.e. taking actions that improve the number but hurt the business. We touched on one example: blasting huge discounts just to spike conversion rate or revenue, while eroding margins and brand equity (congratulations, your conversion is green but you gave away all your profit – not a win). Another example: say you have "average ticket resolution time" as a customer service metric. The team might start closing support tickets quickly to hit the target, but not actually solving customer problems, leading to unhappy customers. In e-commerce, a classic is focusing on ROAS (return on ad spend) or CAC (customer acquisition cost) in a silo – a marketing team might be incented to get a low Cost Per Acquisition (CPA), so they only target super-cheap keywords or remarket to existing customers. CPA plummets (yay, green!), but you're not acquiring *new* customers, or you're cannibalizing organic sales. **To avoid gaming, make sure your KPIs are balanced and tied to true business value.** And keep an eye on the health of the whole business, not just the one metric. This is why I prefer a small set of metrics: it's harder to game *all five* at once without actually improving the business. If someone tries, usually one of the other KPIs will reveal the truth (e.g. margin or repeat rate might go red if you abuse discounts to game conversion). As a leader, foster an honest culture around metrics: if a number is green because of a one-time gimmick or accounting trick, call it out and refocus on sustainable performance. We're after real wins, not just pretty scorecards.

Recognizing these anti-patterns helps you maintain the integrity of your metrics program. The goal is a **culture of transparency and**

continuous improvement. Green means good, red means action –
that's it. No finger-pointing, no sandbagging targets, no vanity moves.
Just a united team, determined to push more metrics into the green
each week the right way.

SCORECARD SUCCESS STORIES: REAL-WORLD WINS

It's easy for me to preach the metrics gospel, but what does it actually achieve in practice? Let me share two brief stories – these are real
e-commerce businesses that transformed their results by embracing the
kind of scorecard and discipline we're talking about.

Cherry Republic's $1.2 Million Save – Cherry Republic is a
Michigan-based D2C brand selling cherry products (jams, candies, you
name it). They implemented an EOS-style scorecard and a weekly leadership huddle to review their five numbers every Tuesday. In just the
first ten weeks, their scorecard lit up a warning that might otherwise
have gone unnoticed. One of their metrics concerned inventory stock-
outs (as a support metric under conversion, they tracked the
percentage of SKUs in stock – something vital for a product-centric
retailer). By week 10, the team saw on the scorecard that **out-of-stock
SKUs had spiked into the red zone**. That was a huge red flag – products were selling faster than production replenishment. Because the
metric was right there in their face, the leadership team immediately
rallied. They reassigned production resources, expedited some supply
orders, and got those popular SKUs back in stock before the holiday
rush. The result? They captured an estimated $1.2 million in Q4 sales
that would have been lost had those items remained unavailable. This
wasn't magic; it was the **power of visibility and fast action**. The scorecard didn't prevent the stock issue from happening, but it ensured the
team caught it early and responded in days, not months. Cherry
Republic's CEO said that before adopting a scorecard, they likely
wouldn't have caught the stockout problem until looking at monthly
or quarterly numbers – by then, the damage would be done. This win
cemented a culture where now every Tuesday, if a metric is red, the
attitude is *"thank you for telling us – now let's fix it immediately."*

The Omnichannel "Deploy Frequency" Turnaround – One of our
clients is a mid-market apparel retailer with both brick-and-mortar
stores and a growing online presence. When we first engaged, their e-

commerce site hadn't seen a meaningful update in months. The IT team was mostly focused on keeping the point-of-sale systems humming for the stores, and the website was treated as "set it and forget it." We convinced the leadership to add **Deployment Frequency** as a Hero KPI on their scorecard, alongside the usual suspects like conversion and traffic. At first, this metric was blood-red: 0 deploys in the last 30 days. It was a glaring indicator that digital innovation had stalled. Seeing that red square on the scoreboard every week created some healthy discomfort. The COO asked the obvious question: "Why aren't we improving our site continuously like our online-only competitors do?" That question led to action – they rebalanced developer resources and made it a goal to have at least one meaningful site update or new feature live every month (target: Green if ≥2 deploys/month). Over the next two quarters, they went from 0 to 3 deploys per month on average. They started small – a new homepage banner here, a checkout flow improvement there – but momentum built. The increased deployment cadence not only improved the site (faster load times, fresher content, smoother mobile UX – all of which helped conversion rate tick up by roughly 15% over six months), it also signaled to the whole company that **e-commerce was no longer on the back burner**. In-store staff started using the improved buy-online-pickup-in-store (BOPIS) features that one deployment introduced, boosting omnichannel sales. The CEO later told me the simple act of tracking "deploys" publicly made the digital team more *innovative* and the rest of the team more supportive of digital initiatives. It united them under a shared goal of making the online experience as dynamic as the in-store experience. The metric eventually "graduated" from red to green and became a point of pride.

These case studies highlight a pattern: when you have the right metrics and a cadence to review them, you catch issues early and create accountability to act. Cherry Republic saved sales by responding to an operational issue surfaced by their scorecard. The apparel retailer kickstarted innovation by measuring it and making it a priority. **What gets measured gets improved** – it's a cliché because it's true. And conversely, what's not measured often gets neglected.

ADAPTING KPIs TO YOUR E-COMMERCE ARCHETYPE

While I'm a huge fan of our five default Hero KPIs, one size doesn't always fit all. Depending on your business model, you might swap out or tweak a metric to better align with what "success" means for you. Let's talk about a few common e-commerce archetypes and how they might adjust the scorecard:

"Sidecar" B2B Ecommerce: This is a scenario where you're primarily a B2B company (with sales reps, catalogs, large POs) but you've added an e-commerce channel as a new arm of the business. In this case, one of the most meaningful metrics might be **New Accounts Created** via the website. Essentially, how many new B2B customers or leads is your online channel bringing in? I would consider swapping out "Qualified Sessions" for something like "New Online Accounts" or "Online Lead Conversions." The reason: in B2B, pure session count is less relevant if those visitors can't purchase without an account. Also, you might track the total number of new accounts across all channels, but use e-commerce influence as a driver (e.g., many accounts might start by browsing online). The hero metric could be the total new accounts opened (with a note that e-commerce should be influencing an increasing share of those). This focuses the team on how e-commerce supports overall customer acquisition in a traditionally offline business.

Omnichannel Retailer: If you have physical stores and an online store, a critical metric is the intersection between the two. A popular choice is **BOPIS (Buy Online, Pick Up In Store) Percentage** – essentially the proportion of online sales or orders that are picked up in store. This metric indicates how well you're blending channels and driving foot traffic via online. You could add BOPIS % as one of the Hero KPIs, or at least a closely watched support metric under conversion or retention. For example, if currently 10% of online orders are picked up in store, you might target 15% as more customers embrace the convenience. Alternatively, measure BOPIS as a percentage of total store sales (e.g., store sales that were influenced by online). The reason this matters: omnichannel customers (those who interact both online and in-store) are often more valuable and loyal. So increasing BOPIS can correlate with higher lifetime value. By tracking it, you encourage initiatives that integrate channels (like online-local inventory visibility, in-store pickup promos, etc.). If I were leading an omnichannel brand, I'd strongly consider making BOPIS % a hero metric – it keeps both ecom and store teams aligned on a united goal of seamless customer experience.

Pure-Play Ecommerce (Consumables Subscription Model): If you're a digitally native vertical brand or any pure-play ecom selling something people buy repeatedly (e.g., coffee subscriptions, pet food, meal kits), then **Subscription Opt-In Rate** might trump a generic repeat purchase rate. This metric would track what percentage of customers (or orders) are enrolled in a subscription or auto-replenishment program. For example, if 30% of your orders are now via subscription, that's a strong indicator of locked-in future revenue. You might swap out the "60-day repeat rate" for "Subscription Enrollment Rate." This focuses the team on converting one-time buyers into subscribers, which is often the lifeblood of consumables businesses. A benchmark here: many good subscription-based brands aim for 20-30% of new customers choosing a subscription option if offered a slight incentive. By tracking it, you underscore the importance of loyalty and convenience for customers. The key is to ensure those subscribers stick (so maybe a secondary metric on churn/cancellation rate is useful for the retention team).

Pure-Play Ecommerce (High AOV or One-time Purchase Goods): Conversely, if you sell big-ticket or infrequently purchased items (high-end furniture, luxury goods, etc.), a 60-day repeat metric might be meaningless (nobody buys a new couch every 60 days). In that case, you might designate a different retention or engagement metric as a Hero KPI. For instance, **Customer Lifetime Value (LTV)** over a longer horizon, or even **Net Promoter Score (NPS)** as a proxy for customer satisfaction if repeat purchase is rare. NPS is a bit of a vanity metric if used wrong, but for luxury brands, tracking changes in NPS could be a vital sign of brand health. You might also track **"First-to-Second Purchase Time"** (how long it takes a customer on average to buy again, even if that's 12-18 months) as your retention KPI. The point is: choose a metric that reflects loyalty or customer advocacy in a way that makes sense for your product's purchase cycle.

The takeaway is **don't be afraid to customize your scorecard to**

your business model – just do it thoughtfully and sparingly. The five metrics should still be the most important things, but what those five are might vary. A B2B-heavy business cares about accounts and reorders, an omnichannel cares about online-offline synergy, a subscription business lives and dies by recurring revenue. Pick the metrics that correspond to your win conditions. And if you're not sure, start with the default five I gave and see where the fit is awkward – that's usually a clue. (For example, if you're tracking repeat rate but 90% of your customers *only ever need one* of your product, then repeat rate isn't a useful North Star; time to swap it out.)

One more note on archetypes: **communicate to your team why you chose these metrics.** If you decide to track BOPIS, explain to store managers how that benefits them (online driving foot traffic). If you swap in NPS, make sure everyone understands how customer satisfaction links to long-term sales. When people see the logic, they're far more invested in hitting the targets. Ultimately, whether you use my five or a variant, the key is keeping the list short, the metrics actionable, and the whole organization aligned behind them.

IMPLEMENTING YOUR SCOREBOARD: TOOLS AND TIPS

By now you might be thinking, "Okay, this sounds great, but how do I actually build this scorecard and track it every week?" The good news is you don't need to spend a fortune or months of development to get a dashboard up. Remember, keep it simple and use tools you already have. Here are some practical ways to implement your scoreboard:

Looker Studio (Google Data Studio): This free tool from Google is fantastic for creating a one-page dashboard connected to sources like Google Analytics, Google Sheets, or BigQuery. You could pull in your five metrics from GA (for things like sessions, conversion, AOV) and maybe a Google Sheet or database for something like deployment count or NPS scores. I've built scorecards in Looker Studio that update automatically and are easily shareable via a link. It's drag-and-drop, and there are templates out there – you don't have to be a BI expert or pay for a BI tool.

ShopifyQL or Native Platform Tools: If you're on Shopify (Plus), the ShopifyQL Notebooks or even the basic dashboard might suffice for a few metrics. Shopify's analytics can show conversion, AOV, etc., and you can augment with a custom app or script to feed something like deploy count. Adobe Commerce (Magento) users often have Adobe Commerce Business Intelligence (formerly RJMetrics) which is very powerful – you can set up a dashboard there with all your KPIs. The point is, leverage whatever analytics platform comes with your e-commerce system, if it has one, and customize a simple report. Most platforms let you schedule an email or export, so you could even get a weekly email every Monday with the five numbers and their status.

Plain Old Spreadsheets: Don't laugh, but a Google Sheet or Excel sheet manually updated might be the easiest starting point, especially as you're ironing out definitions. I've had clients simply maintain a Google Sheet where each column is a week and each row is a KPI. They color the cells green/yellow/red by hand based on the thresholds. Takes maybe 15 minutes a week to update, and you'd be amazed how many insights you can get from just plugging those numbers in yourself (it forces you to look at them). You can always automate later; first prove that the habit and focus add value.

EOS Software: Since we've referenced EOS, I should mention there are dedicated EOS software tools (like Ninety.io, Traction Tools, and others) that include scorecard modules. If your company is fully on EOS and uses one of these, by all means use the built-in scorecard feature. They often have nice meeting agendas and IDS (Issue–Discuss–Solve) tracking integrated too. Just be cautious: don't twist your metric choices to fit the software. If it's cumbersome to input or view certain metrics, consider whether the tool is helping or if a simpler solution would be better. The tool is just a means; a basic solution that everyone actually uses beats a fancy one that people ignore.

Whichever tool you choose, enforce the **one-page, one-screen rule**. Configure your dashboard to show all five metrics at once without scrolling. Put it up on a big screen in your office or share your screen in virtual meetings. It should be literally the "scoreboard" on the wall. Also, try to show trends in a minimalist way – like a small line sparkline or a comparison to last week – but avoid the temptation to add 10 extra charts. If someone wants to dive deeper into, say, why conversion moved, they can go into Google Analytics outside the scorecard meeting. The scoreboard itself stays high-level.

Another tip: set up **automated alerts** if possible for red conditions. Many analytics tools or even Google Sheets scripts can trigger an email or Slack alert if a number falls below a threshold. For example, if weekly conversion drops below the red line, an alert could notify the KPI owner immediately ("Hey, conversion is 2.4% this week, below 2.7% red threshold"). This way, the response can start even before the official meeting. Just be careful not to create alert fatigue – only set up alerts for truly critical, time-sensitive drops.

Finally, maintain data hygiene and clarity. Make sure everyone agrees on the exact metric definitions and data sources. If "qualified session" is a custom Google Analytics segment, document it so folks know. If "repeat rate" is measured by new customers from 2 months ago who bought again, spell that out. One of the worst things is disputing numbers in the meeting ("I have a different figure in my report"). Avoid that by getting consensus on the data source of truth for each KPI. It might take a little setup work with your analytics team or agency, but it's worth it.

The bottom line: **don't over-engineer this.** The value lies in using the scoreboard, not in perfecting the dashboard aesthetics. A scrappy dashboard that your team actually looks at weekly beats a super-polished dashboard that sits forgotten. So get something in place quickly using available tools, and refine as you go. The goal is to start seeing those weekly colors and trends, and to start having the right conversations.

THE 15-MINUTE WEEKLY SCORECARD HUDDLE

Now that your metrics are defined, owners assigned, thresholds set, and dashboard ready, it's time to actually run the process that ties it all together: the weekly scorecard meeting. This is the **heartbeat** of your metrics-driven culture. And the good news – it's only 15 minutes a week. (If that sounds too short, trust me, when you're focused it's plenty.)

Here's how to run a tight weekly scorecard huddle:

Same Time, Same Day, Every Week: Consistency is key. For example, *every Tuesday at 11:00 AM* the leadership team gathers (in-person or

Zoom) for the scorecard review. Make it a recurring meeting that everyone knows is sacred. Starting on time is crucial – if you wait for stragglers, it balloons. Treat it like a stand-up meeting; promptness sets the tone.

Screen Share the Scoreboard: If remote, one person pulls up the scorecard dashboard and shares their screen. If in person, project it on the TV or have printed one-pagers. Everyone should be looking at the same five numbers, updated through the end of the previous week. No PowerPoint, no fluff – just the raw scoreboard.

Run Through the Numbers Top to Bottom: The moderator (could be you as the e-commerce lead, or whoever owns the meeting cadence) simply reads each metric and its status. "Qualified Sessions: 52,000 – that's green (up from 50k, above our 51k target). Mobile Conversion: 2.6%, that's yellow (down from 2.8% last week, below 2.7% threshold for red... etc.)." Do this for all five. **No deep discussions here**, just reporting the score. Think of it like a sports announcer giving the play-by-play.

Flag Yellows and Reds – Assign Actions: As you hit a metric that's yellow or red, briefly pause and make sure there's an owner taking responsibility. Since you pre-assigned owners, it should go like: "Mobile Conversion is yellow at 2.6%. Jane, as the owner, you're on point – what's the status?" Jane might reply with one sentence: "Yes, we're investigating a possible bug in the cart on mobile since Friday, team is on it." Great – the facilitator can note that and perhaps tag it for follow-up in a separate problem-solving meeting. If something is red and no one has a theory yet, that's okay; assign someone (usually the owner) to dig in right after this meeting and report back. In EOS, truly pressing issues would go to an "IDS" (Identify-Discuss-Solve) session in the longer weekly Level 10 meeting. If you're not doing EOS, you can simply schedule a quick deep-dive meeting with the relevant folks. **The key is not to solve it in the scorecard meeting.** Just acknowledge it and assign it.

No Excuses, No Debates: This is perhaps the hardest part – keeping the meeting strictly factual and action-oriented. If a number is red, it's red. I don't allow a lot of "oh, but this number is usually low in summer, maybe we should re-color it" or "traffic was down because the email promo was delayed, so it's not *our* fault." Save all that for analysis later. In the huddle, red means *figure it out*, not *explain it away*. By cutting off lengthy discussion, you ensure the meeting stays 15 minutes. If someone raises a valid point ("our tracking might be off"), note it as an action item to investigate offline. Trust me, this discipline is liberating – meetings stop dragging on because everyone knows the rules: report the number, mark issues for follow-up, move on.

Celebrate the Greens: Don't gloss over the good news! If a metric went green after being yellow, give a quick virtual high-five or kudos to the owner/team. "Email capture rate is green for the first time – awesome work, team!" A little recognition goes a long way. It also reinforces positive behaviors. We sometimes take the wins for granted and only focus on problems, but a great culture is both *determined* to fix problems **and** *quick to celebrate*improvements. It keeps morale up and momentum going.

End on Time (or Early): When you finish the last metric, and any quick assignments for yellows/reds, conclude the meeting. Typically this should be around the 15-minute mark. If you're consistently running over, check if someone is dragging into analysis or off-topic. Cut that out. I personally set a timer for 15 minutes when we start – it sounds a bit strict, but it works. After a few weeks, people get into the rhythm and you won't even need the timer. Ending on time shows respect for everyone's schedule and reinforces that this is a focused huddle, not a meandering discussion.

After the meeting, the real work happens: teams execute on the action items for any metric that needs love, whether it's a deep-dive analysis, an A/B test, a firefight fix, or a new idea to try. Those actions might be handled in separate working sessions. The beauty is that the scorecard has surfaced *exactly* where to spend your energy that week. No more guessing or going by gut alone – you have a data-driven target for improvement.

One more pro-tip: hold this meeting even when you've had a bad week. Especially then! Sometimes I see teams cancel the scorecard meeting if they know numbers are down ("we had a site outage, everything will be red, let's skip this week"). Nope, that's when the huddle is most important. You need to reinforce the culture that facing the music is non-negotiable. By confronting the ugly weeks, you also get the chance to quickly strategize fixes and reassure the team that a plan is in place. And when things return to green, it feels that much better.

In my experience, this 15-minute ritual becomes the *pulse* of a truly data-driven organization. It keeps everyone aligned, accountable, and agile. I've even had team members tell me they start to love scorecard time, because it gives immediate clarity – a sense of "here's how we did, here's where we go next." When you reach that stage, you know the scoreboard mentality has sunk in deep.

THE 90-DAY KPI RESET: A PLAYBOOK FOR ONGOING FOCUS

Implementing a new scorecard is a big change, and it can be tempting to fall back into old habits over time (like adding more metrics or losing the weekly discipline). To guard against that, I recommend running a **90-Day KPI Reset plan** – essentially a quarterly cycle to keep your metrics program lean, mean, and continuously improving. Here's the playbook:

Step 1: Purge and Prioritize (Day 0): Start by purging any existing metrics clutter. Archive those massive dashboards that look like a spaceship control panel. Remove or hide any KPI that hasn't directly driven an action in the last 60 days. This is a cleansing process – get back to a blank slate except for the essentials. Then, from that clean state, prioritize your new Hero KPIs (the five we've been discussing). Write them down clearly with definitions, owners, and why each matters. This "lock-in" moment is important – you're making a contract with yourself and the team: *for the next 90 days, these are our focus metrics, period.*

Step 2: Baseline and Target (Week 1): As you kick off the quarter (or whatever 90-day period), establish your baseline for each KPI (what is it today or on average recently?) and set a concrete 90-day target. For instance, "Current mobile conversion is 2.5%. By the end of Q, we want 3.0%." Make sure the target is realistic yet ambitious – typically 10-15% improvement for a metric in a quarter is a healthy challenge, though it varies. Also define the Red/Yellow/Green ranges now if you haven't already. Document all this in your scorecard sheet or dashboard so it's visible. This gives you a clear finish line to run toward and helps rally the team. It's also a way to measure success of this whole process.

Step 3: Weekly Huddles and Experiments (Weeks 1–12): For the next 12 weeks, you religiously run the weekly 15-minute scorecard huddles as described. After each huddle, for any KPI that is yellow/red or not trending toward the target, design a small experiment or action for the upcoming week. For example, if qualified sessions are lagging, the marketing team might run a new campaign or test a new channel that week. If AOV is flat, maybe test a different upsell on the site or a limited-time bundle offer. These don't have to be massive projects – think of them as weekly sprints or tweaks guided by the score. Then the following week, see if there's movement. This creates a rapid feedback loop: every week you try something and see the impact on the number. It turns the scorecard into a living tool for experimentation, not just observation. Keep a quick log of these experiments somewhere; it's great for end-of-quarter reflection.

Step 4: Celebrate Wins and "Graduate" Stable KPIs (Day 90): At the end of the 90 days, hopefully you'll be looking at a much greener scorecard and ideally a few of those targets met or exceeded. Take time to celebrate that with the team – seriously, call it out in a meeting, give kudos, maybe send the team some swag or gift cards if a major goal was hit. Positive reinforcement is huge in building a lasting culture. Now evaluate each KPI's journey. If a metric has been consistently green for, say, 4+ weeks and you feel it's under control, you might consider "graduating" it off the active weekly scorecard for the next cycle. For example, maybe AOV was a big focus and went from $75 to $85 and is holding steady – great, that might graduate. "Graduating" means you'll still monitor it via an automated monthly report or alert for any dips, but it no longer needs to be one of the five you discuss every week. This frees up a slot to introduce a new Hero KPI for the next 90 days. Perhaps now you're ready to tackle a new challenge, like improving customer service response time or increasing new customer acquisition – whatever emerged as the next opportunity.

Step 5: Rinse and Repeat: Design your next 90-day scorecard cycle. This might carry over some metrics that still need work (anything still yellow/red by day 90 likely stays on the scorecard), and bring in one or two new ones if some graduated. The process then repeats: set new targets, run the weekly cadence, drive improvements. This iterative approach prevents stagnation. It also addresses the worry "but what about [metric X] that we're not tracking?" – don't worry, if it becomes mission-critical, it can be rotated into focus in a future cycle. Over the course of a year, you might tackle 7-8 different metrics deeply, just not all at the same time.

The 90-day reset plan ensures you never slip back into the vanity metric swamp. By limiting the time horizon, it creates urgency ("we have 3 months to move this needle") and then a natural checkpoint to evaluate and adjust. It also aligns nicely with quarterly business planning or EOS "Rocks" if you use those. And by planning to remove some metrics after success, you avoid the scorecard bloat. I've seen companies a year into this process that have entirely different scorecards than when they started – and that's a good thing because their business evolved and their focus shifted to new growth levers. Yet they never tracked more than five at a time.

One pro tip for post-graduation: set up **monitoring/alerting** for any metric you take off the weekly list. For instance, if AOV is no longer a weekly KPI, have an automated monthly report or threshold alert so you'll know if it starts dropping significantly. That way you can catch any backslide in a graduated metric and potentially put it back on the scorecard if needed. But chances are, once you've built good habits and improvements around a metric, it tends to sustain (especially if you've addressed root causes).

In 90 days, you can completely change your company's trajectory by focusing on what matters and empowering your team to act on data. And quarter by quarter, you build a **sustainable growth flywheel**: pick the right metrics → drive improvements → institutionalize gains → move on to the next high-impact area. That's the essence of moving from vanity to value in a repeatable way.

Onward: Metrics Today, Accountability Tomorrow

We've covered a lot in this chapter: from gut-punch wake-up calls about over-tracking, to building a focused five-metric scorecard, to instilling a weekly discipline and even rethinking your KPIs every 90 days. By now, you should have a clear game plan to transform your organization's approach to metrics – turning data into a true performance *scoreboard* that drives action every week. The end result? A team that is **determined** (no more hiding from red numbers), **united** (rallying around the same goals), and **innovative** (constantly testing ways to move the needle).

Before we close, let me emphasize one more point: this scoreboard mindset isn't just for your internal team. It extends to every partner you work with. If you hire an agency for SEO or PPC, or you use a vendor for email marketing, they too should be accountable to these metrics that matter. In the next chapter, we'll dive into how to hold your **agencies and vendors accountable** to results, using the very scorecard and KPIs you've now established. After all, your playbook for growth should be shared by everyone playing on your team – in-house or external. By setting clear expectations and tying vendor performance to your Hero KPIs, you ensure that all the brilliant strategies and execution we've covered in this chapter don't get undone by misaligned partners.

You've built the scoreboard and started playing the game to win. Now, let's make sure every player (internal or external) is keeping score the same way and striving for those same wins. **On to Chapter 4.4, where we turn our focus to aligning and challenging your external partners with this metrics-driven approach.** Get ready to extend your culture of accountability beyond your four walls – because true e-commerce growth requires every single contributor to bring their best, with eyes on the right scoreboard. Let's keep this momentum going!

4.4: PARTNER SELECTION & VENDOR CONTROL

"*The contract's signed – time to kick back, right?*" I hear this a lot, and it always makes me shake my head. Too many eCommerce leaders pour energy into selecting a platform or agency, then breathe a sigh of relief as soon as the ink dries. In reality, signing the deal is just the beginning. The real work starts **after** you've picked your eCommerce partners. In this chapter, I'll share how to choose the right partner **and** how to actively manage that relationship to drive growth. My goal is to challenge your assumptions and give you practical steps to get more from your eCommerce vendors (and to know when it's time to walk away).

Let's start with a story I see all too often: An eCommerce director partners with a reputable agency to build their online store. The first year goes great – lots of new features rolling out, good momentum. But fast-forward a couple of years, and things have *drifted*. The budget's the same, yet the output has slowed to a crawl. One brand I worked with discovered that over three years with their previous agency, their development velocity **halved** while spending stayed constant. Their backlog kept ballooning, projects stalled, and frustration mounted on both sides. What happened? In my experience, this is a classic case of **"vendor drift."** Small issues – unclear scope, staff turnover, complacency with KPIs – accumulate like fog around the project. Without the right controls, that once-promising partnership can devolve into a costly stalemate.

How do we prevent this? By **selecting the right partner up front and then governing the relationship** deliberately. We'll cover both parts here. First, I'll walk through choosing an eCommerce agency or provider that truly fits your business. Then, I'll dive into how to set expectations, maintain control of your tech **and** your timeline, and keep your vendor relationship healthy and accountable. Along the way, I'll share some anonymized case studies from my years at Creatuity and beyond – real stories of what works and what doesn't. By the end, you should have a game plan to turn your vendors into true partners in growth. And if they can't get with the program, you'll be ready to show them the door. Sound good? Let's get into it.

CHOOSING A PARTNER, NOT JUST A VENDOR

Successful eCommerce growth starts with **picking the right partners**. I say "partners" because you want more than a vendor who just executes tasks for a fee. You want a team that *invests* in your success, challenges you when needed, and sticks around for the long haul. So how do you find an agency or platform provider that fits that bill? Here are a few key factors I always look for, and some pitfalls to avoid:

Relevant Experience and Industry Expertise: Look for a partner who understands *your* market or business model. This doesn't mean they must have built an store exactly like yours (that might be impossible), but they should have experience in similar terrain. For example, at my agency (Creatuity), we've worked a lot with farm-and-home retail chains – a niche of omnichannel retail. When a new farm & home merchant comes to us, we already grasp things like huge SKU catalogs and in-store pickup nuances. Similarly, some agencies excel at B2B eCommerce with complex catalogs, or at fashion D2C with fast product cycles. One mid-market B2B retailer I know struggled with a massive product catalog and poor search on their site. They switched to an agency that specialized in Product Information Management integrations and enterprise search. Within a couple of months, their catalog errors dropped dramatically and conversion on search results jumped – because the agency *knew how to solve that specific problem.* The bottom line: **don't hire a generalist** if you have specific challenges. An agency might have glowing accolades for pretty D2C fashion sites, but if you're a B2B industrial supplier with 100,000 SKUs, that likely isn't the right fit. Find a specialist who has solved your kind of pain point before.

Technical Expertise (Even on "Easy" Platforms): I have a bit of a technical background, so I'm biased, but I rank technical chops very highly. Before you get wowed by fancy design portfolios, make sure the agency truly knows the eCommerce platform or technology stack you're using (or considering). This is critical whether you're on an enterprise platform like Adobe Commerce or Salesforce, an up-and-coming platform like Shopware or Medusa, or a SaaS platform like Shopify/BigCommerce. Don't assume a SaaS platform means you can ignore tech skills – I've seen "fully hosted" stores brought down by agencies making coding mistakes or misconfiguring something basic. A quick test: ask what certifications their developers have, or get a little geeky and inquire about a tough technical challenge they solved on a recent project. The goal isn't to stump them with trivia; it's to gauge depth. For instance, ask: *"What's the hardest problem you've solved on Platform X, and how did you approach it?"* The answer will tell you a lot. I love hearing war stories about creative debugging or scaling fixes. If they can't share any, or they only speak in buzzwords, that's a red flag. Another red flag: an agency that *only* ever proposes plug-ins or apps for everything. I recently consulted with a retailer whose agency kept installing app after app to avoid custom development. They ended up with a bloated site that still didn't meet their needs, because the agency simply didn't have the skill to write custom code. The right partner will know when to use an off-the-shelf module versus when to build a tailored solution, and they won't shy away from either approach.

Stability and Scalability Track Record: Here's a scenario you want to avoid – your site is booming and traffic is spiking, but **every time you run a big promotion, something crashes**. If an agency or platform can't keep your site stable on the biggest days of the year, that will absolutely limit your growth. So ask potential partners how they handle **scaling and performance**. Do they do load testing? Capacity planning for events like Black Friday? Have they worked with sites that get sudden traffic bursts from, say, viral campaigns or TV spots? I've worked with brands that got featured on national TV – imagine going from baseline to 5× traffic in minutes. One client was planning a spot on a morning show, and we spent weeks beforehand re-architecting their site for auto-scaling, tuning the database, even implementing a queue system so the site wouldn't buckle under the rush. It paid off with 100% uptime during the event and record sales. When interviewing agencies, listen for that kind of foresight. If you mention Black Friday and they *don't* immediately talk about preparation, that's worrisome. As I often tell merchants: everyone's site does NOT have to go down on Black Friday – if yours does, something's wrong. It could mean you've outgrown your platform or your agency's skillset. Either way, a good partner should be proactive about stability. Ask how they monitor uptime and respond to incidents. A competent team will almost *welcome* this question and have a confident answer like, "Our goal is that it **never** crashes – here's how we architect and what we do if something goes wrong".

Cultural Fit & Communication: This is huge and often overlooked. Your agency's culture needs to mesh with yours. If it doesn't, even a technically skilled, well-priced vendor relationship can turn toxic. I've seen a real-life example of this: a Texas-based company hired a big-name New York agency (paying top dollar) and lived to regret it. The culture clash was palpable. As soon as any stress hit the project, it became **"those rude New Yorkers"** vs **"those slow Dallas folks,"** and trust evaporated. The engagement fell apart despite both sides being competent – simply because they couldn't communicate without friction. Cultural fit can mean regional attitudes (as in that case of North Texas informality clashing with NYC directness), but it's also about **company values**. Are you a fast-moving, "get it done yesterday" organization? Then an agency that values methodical long-term planning might frustrate you (and vice versa). Do you expect a friendly, personal relationship or do you prefer strictly business? Ensure the agency's style aligns with yours. Remember, if things go well, this partnership could last 5+ years – you better actually enjoy working with them. During the vetting process, pay attention not just to what they say but *how* they say it. Are they listening to you? Do they communicate clearly? For example, if you prefer phone calls but they seem to do everything by email and Slack, clarify that upfront. One more tip: ask for client references and *actually call them*. Then ask those references about the *day-to-day experience* of working with the agency. Don't just ask "Were you happy with the final project?" Dig into communication, responsiveness, and whether the agency felt like an extension of their team. This "cultural due diligence" can save you a lot of pain down the road.

Value vs. Cost (Avoiding the Low-Bid Trap): Budget matters – I'm not going to pretend it doesn't. But if your entire selection process is "whoever comes in cheapest wins," you're setting yourself up for trouble. I've seen companies line up three proposals and automatically pick the lowest bid, or toss out the high and low and take the mid-range, as if they're buying a commodity. That approach **ignores quality** and can backfire spectacularly. The cheapest proposal often results in endless change orders and budget overruns later. Some agencies deliberately underbid and then nickel-and-dime you on scope changes – you end up paying as much as the higher bid would have been, but with more headaches. The other risk with bargain-basement agencies is they may staff your project with junior freelancers rotating in and out. You lose continuity, things move slower, and quality suffers. On the flip side, the most expensive option isn't automatically the best either – you might just be paying for a fancy brand name. I had a client switch to us after using one of the priciest agencies in the country; even though money was no object for them, the engagement failed because, again, culture and values were mismatched. My advice: **evaluate total value**, not just sticker price. Consider the agency's expertise, their integrity (are they upfront about what's included?), and the potential *cost of things going wrong* with an unproven low bidder. Also, be open about your budget range when asked. I know, I know – in negotiations "he who speaks the number first loses," right? But I don't buy that. In a trust-based relationship, transparency wins. If you have a top-shelf champagne vision but a bargain beer budget, it's better to tell the agency on day one than to hide the budget and get proposals that you can't afford. A good agency will work with you to scope realistically or break the project into phases that fit your budget. If you tell me your budget is $X, I can say, "All right, with $X, here's what we can do." Then *you* decide if that's acceptable or if you need to adjust your goals (or find more budget). This honest conversation up front saves everyone time and prevents disappointment later. The same goes for timeline – we'll talk more about realistic timelines soon, but be wary of any firm that promises the moon overnight at a bargain price. In eCommerce projects, as in life, if it sounds too good to be true, it probably is.

Integrity and the "Trust Factor": Finally, gauge their integrity. Are they asking you tough questions about your business, or just saying yes to everything? An agency that *never* pushes back, that promises every feature you want without discussing trade-offs – that's a red flag. You actually want a partner who will say "no" (or at least "let's rethink this") when you propose something that isn't in your best interest. I often tell merchants to think of your agency like a good friend. You don't want a friend who just flatters you; you want one who stops you from making a fool of yourself. In the same way, a great agency partner will tell you if a certain customization will hurt site performance, or if your timeline is unrealistic, or if they think a cheaper app can solve your problem instead of custom dev. That honesty is gold. During vetting, ask them about a time they disagreed with a client's idea – and how they handled it. Their answer will reveal a lot about their **professional honesty**. And as a client, be prepared to hear some answers you might not love. It's okay – that's why you're hiring expert partners. If an agency is brave enough to (politely) say "we recommend against that approach" and explain why, you likely have a keeper. By contrast, steer clear of the "whatever you say, we do" yes-men agencies. As one of my colleagues colorfully puts it, there's the "*If the man wants a green coat, turn on the green light*" types – those who will promise anything to get the deal. That's not partnership. That's a vendor happy to collect your checks until the project blows up. **Trust but verify**: ask for examples, talk to references, and listen for candor. The agency that admits "We're not the best fit for that one piece, but we excel at these others" or "Client X left us because we just weren't aligned on strategy" – that honesty is refreshing and usually means they're the real deal. In short, you want a relationship that's built to go **far, not just fast**.

Before we move on, let me emphasize a *critical* piece of fine print: **Ownership of your code and data.** This is non-negotiable in any vendor contract. I've been beating this drum for decades: you, the merchant, must own your website code (unless it's on a SaaS platform where owning code isn't applicable) and you must own your data. Never sign an agreement that gives the agency any sort of exclusive rights over code that you paid to have developed. It's normal for agencies to retain ownership of their *generic* reusable components, but anything specific to your site should be yours. And yet, I still encounter businesses who don't even have access to their own code repository because of a sneaky contract clause. In one case, an eCommerce director told me their agency refused to give them *view* access to the codebase – the company was essentially flying blind, unable to even see the code running their store. That's outrageous. If you end up in that situation, your agency owns **you**. They could hold your site hostage if disagreements arise. So read the fine print and push back on any terms that restrict your ownership or access. The same goes for data exports, admin accounts, etc. If it's your business, you should control it. In fact, let's expand on this in the next section – because maintaining control is just as important after you've picked the partner as it is during selection.

Setting the Foundation and Keeping Control

Congratulations – you've chosen an eCommerce agency or technology vendor. Now the real work begins: **establishing how you'll work together and making sure you stay in the driver's seat.** A great partner won't be threatened by this; in fact they'll welcome a proactive, engaged client. Here's how to set the stage from Day One of the engagement:

Kickoff with Clear Expectations (Roles, Goals, Responsibilities). The first serious meeting with your new agency (often called a kickoff) is crucial. This is where you align on project scope, timeline, and how you'll collaborate. Don't let it be a one-sided presentation where the agency just shows a glossy project plan and everyone nods. Instead, expect *and insist* that your team's responsibilities are discussed too. For example, who on your side will provide assets or content? Who will review and approve deliverables? Who will do UAT (User Acceptance Testing) for new features? Many merchants assume the agency "handles everything," but in practice, your internal involvement is key to success. I've seen clients assume they only need to show up for a weekly status call, only to realize late in the game that no one scheduled on their side to do critical testing or data prep – and that can blow up a timeline. A good project manager will ask you about your team's **resources and skills** up front. For instance, *"Who will be our point person for content changes? Do you have an in-house developer to assist or will you rely entirely on us? Who signs off that a feature is working as intended in your business workflow?"* By hashing this out, you might discover you need to assign (or even hire) someone on your end, or adjust the timeline because your one QA or UAT person is stretched thin. It's much better to surface those realities at kickoff than to realize too late that *"oh, we never tested that checkout flow with our ERP"*. Also, be upfront about any known constraints or "ugly" parts of your current state. If your internal system is a mess, or you know your team tends to be slow in giving feedback, say so. The agency can then plan around it or help solve it. We're all adults here with the same goal – launching a successful project. Honesty and clarity at kickoff build the foundation for trust.

Along with roles and responsibilities, use the kickoff to articulate the deeper **goals** of the project beyond just "launch on time and on budget." Certainly, on-time/on-budget is important. But ask your stakeholders (and yourself): *what will make this project a true win?* Sometimes the answer surprises you. I've had clients whose top priority wasn't timeline at all – it was to restore trust after a prior bad agency experience. One client explicitly told us, "My main goal is to not feel like I have to scrutinize every invoice for overbilling". They'd been burned by lack of transparency before, and that anxiety was their big pain point. Knowing that, our team put extra emphasis on regular budget reporting and giving them direct access to our time tracking. We made sure they **never** felt blindsided by a bill. The lesson: if something *other* than the obvious deliverables is profoundly important to you, voice it. Whether it's avoiding downtime, achieving a certain site speed, enabling your team to manage content without IT – whatever your "it really hurts here" issue is, tell your agency upfront. A good partner will take that seriously and adjust how they work to hit that target. This also helps later when tough decisions arise – both sides know what the north star is.

Treat Your Agency as Part of the Team (and vice versa). This is a mindset thing. If you treat your agency like a transactional vendor ("here's a list of tasks, get 'em done"), you're not going to get the best out of them. And if the agency treats you like just a revenue source, they won't get the best out of you either. The modern approach – which we practice – is to blend teams. We consider the client **part of our team** and hope to be seen as part of the client's team. That means building *real* human relationships, not just sending tickets into a void. It means mutual respect: you acknowledge the agency's expertise, and they acknowledge you know your business best. When an agency project manager says "we're in this together" and means it, magical things happen. Trust forms. And trust is like rocket fuel for eCommerce projects – it gets you through the inevitable challenges. One of our project managers put it nicely: the hardest, most complex projects *succeeded* because we had strong trust with the client. How to build that trust? Communication and transparency, for starters. Share not just the *what*, but the *why*. If you're worried about something, say so. If the agency is concerned about a risk, they should voice it. We maintain an open **risk register** that both our team and the client team contribute to and can see in real time. When something goes wrong (and something always will), address it together instead of finger-pointing. It's very much "us vs. the problem," not "us vs. them."

One concrete practice: invite the agency to your internal strategy meetings when appropriate, and have them invite you to their planning sessions for your project. Blurring that line a bit ensures everyone shares context and feels accountable to the same goals. Also, don't keep your agency at arm's length with information. I've seen clients who treat their agency like an adversary – hiding budgets, not sharing true feedback – and then wonder why the relationship is strained. That's like going to a doctor but not telling them where it hurts! As I said earlier, show them where it hurts. If both sides commit to transparency, you can solve problems faster and catch issues before they fester.

Establish Governance Rhythms (QBRs & Scorecards). Earlier I described "vendor drift," where an engagement slowly loses momentum or alignment. The antidote to drift is **regular, structured check-ins at the executive level**. At Creatuity, we insist on Quarterly Business Reviews (QBRs) with each major client. The point is to have a recurring high-level meeting *outside the day-to-day project calls* to evaluate the health of the partnership. I recommend a standing agenda that covers four key areas:

Financials: How are we doing against budget? Are we burning hours or dollars faster than anticipated, or are we under-running? No one likes surprises here, so surface any budget issues openly. Also review any change orders or scope changes in the last period. If there have been a lot, discuss why. I'm not talking about a blame game; it's about identifying patterns. Lots of change requests could indicate the agency underestimated work *or* that your team kept changing direction mid-stream. Either way, it's something to learn from so you can course-correct. Finally, look at any costs beyond the agency: third-party licenses, plugins, hosting, etc. Make sure you're not paying for things you no longer need – you'd be surprised how often a company keeps a redundant software subscription around. We've caught clients double-paying for services because they forgot to cancel an old contract after a new platform provided the feature natively. A quarterly cleanup can save you money.

Delivery & Performance: This is the retrospective on what the agency delivered in the last quarter. We look at metrics like development **velocity** – for example, story points completed or number of tickets closed – whatever measures output in your process. Is the team delivering more, less, or about the same as before? Track it over time. If velocity is dropping (as in that vendor drift example), raise a flag and ask why. We also examine **cycle time**: how long does it take a new idea or request from your business to get live on the site? Is that timeline acceptable, and is it improving or worsening? Slow cycle times might mean bottlenecks we need to address. Another key metric: **quality**, often measured in bug counts or error rates. How many issues are making it to production and then requiring hotfixes? A trend of increasing bugs could indicate rushed work or insufficient testing. Conversely, if quality is high – celebrate that! Essentially, in the delivery section both sides should come prepared with data on how the partnership is performing. Use a shared **scorecard** to track these KPIs over time. I like to use a red/yellow/green system for easy scanning: e.g., if we completed 90% or more of the planned story points, that's green; 70–89% is yellow; below that is red. Or set thresholds for bug counts (we might say 0–2 bugs a month = green, 3–5 = yellow, more is red). The exact numbers will depend on your situation, but the idea is to agree on targets and objectively see how we're doing. And if something is red for two quarters running, it's definitely an item to address (maybe the scope is too ambitious, or the agency needs to add resources, or you need to improve your internal processes). When everyone sees the same scorecard, it brings accountability and focus. It's no longer about "I feel like you guys aren't delivering enough" – you have data to discuss, which is much healthier.

Roadmap & Next Quarter Planning: After looking back, turn the focus forward. Align on the priorities for the next quarter. What new features or projects are coming up? Are there any looming risks or technical debt that need attention? This is where you hash out and agree, at a leadership level, on what the agency will be working on next. It ensures both sides are on the same page and can secure resources as needed. I've sat in QBRs where a client's executive says, "Actually, Project X got deprioritized by our board, we need to pivot to Project Y next month." Imagine if that didn't come up until the sprint planning the week before – lots of wasted effort. The roadmap discussion in a QBR prevents surprises. It's also a chance for the agency to suggest optimizations or upgrades you might not have considered. Maybe they've noticed your mobile site is lagging and propose a performance overhaul next quarter. These strategic conversations elevate the relationship from just ticket-based interactions to true collaboration.

Growth and Business Outcomes: Finally, tie the eCommerce work to business results. Look at metrics like conversion rate, average order value, customer lifetime value, repeat purchase rate – whatever key numbers you're aiming to improve. Is your eCommerce channel growing as expected? If not, why not, and how can the roadmap address that? If yes, how do we pour more fuel on the fire? In our QBRs, we often review the client's overall eCommerce scorecard (some use the EOS model scorecard, for example) to see how things trend over the quarter. This grounds the tech and agency discussion in what ultimately matters: revenue, profit, customer satisfaction. It prevents the agency from operating in a silo. If the numbers are good, both teams can high-five and identify what contributed to that success. If numbers are down, it's a joint problem to solve. The agency might not drive your marketing and merchandising, but they can impact site performance, user experience, and more – so they absolutely should care about your business outcomes. A great partner will treat your KPIs like their own.

After covering these four areas, the QBR should end with clear action items (just like any good meeting). If the scorecard showed some red areas, what's the plan to get them to yellow or green? If new priorities surfaced, does scope or budget need adjusting? The beauty of a regular review is you catch drifting performance early – after one bad quarter, not a full year of mediocrity. And if the agency knows this meeting is coming, they tend to stay on their toes throughout the quarter. It's an accountability mechanism that *both* sides benefit from. In fact, I recommend that you as the client also hold yourself accountable in these meetings. For instance, if you committed to delivering new product data or marketing content by a certain date and you didn't, own up to that. The spirit is not to scold, but to continually improve how client and agency work together.

Own Your "Digital Keys": licenses, logins, and assets. This one might save your business someday: **always keep ownership of your core technology assets.** Buy your own software licenses, register services in your name, and have your own admin access to everything from day one. I've seen agencies offer "We'll purchase that extension for you under our account to make it easier" – politely decline. Even if it's slightly more hassle for you to run the purchase through your procurement, do it. Why? Because I've dealt with too many companies who, upon parting ways with an agency, realized they didn't have the licenses or account control for critical parts of their site. One company spent thousands of dollars to **re-purchase extensions** they already paid for once, just because their former agency had bought those plugins under the agency's account and wouldn't (or couldn't) transfer the license. In another case (a worst-case scenario), an agency had "borrowed" an extension from one client's project to use on another's to save money – a blatant violation of licensing. When the client moved to us, they discovered they weren't legally licensed for a module their site relied on. Yikes. Don't risk these situations.

The same rule goes for hosting and domains. **Host your site in an account you control.** It's fine to let the agency set up or manage the environment, but it should ultimately be *your* account with the hosting provider, or at least you have full admin rights on the server. If an agency ever tells you "No, we keep the server access and we can't give you root/admin," that is a huge red flag. I'd say walk away right then – that's not a partnership, that's a hostage situation. If they won't give you the keys to your own kingdom, they're not a partner you want. Your eCommerce site is part of your business's lifeblood; you absolutely need the ability to access and control it if push comes to shove.

If you're on a SaaS platform (like a Shopify, BigCommerce, etc.), you might not have a server login, but you still should have the master admin account to your store and any apps. Make sure **your company's email** (not the agency's) is the primary owner on these services. I've seen cases where the eCommerce manager left a company and no one else knew the login to their platform or analytics – chaos ensues. Don't let that happen. **Document the accesses** and ensure at least two people on your team can get into everything critical. If someone leaves, rotate those passwords or keys as part of off-boarding.

One more often-overlooked asset: your code repository. If the agency is writing custom code for you, insist it lives in a repository where you (and your future partners) can get it. Many agencies use GitHub or Bitbucket – have them create a repo under *your* organization, or at least give you full access rights to their repo. That way if you part ways, you don't have to beg for a code dump; you already have the full history and codebase. Also consider stipulating in the contract that you get copies of any important documentation or configuration. I

know it's tedious to think about the end at the beginning of a relationship, but having these ducks in a row turns a potential messy breakup
into a relatively smooth transition if it ever comes. It's like insurance –
hope you never need it, but you will be glad to have it if you do.

Manage Scope, Budget & Timeline Like a Three-Legged Stool.
Let's talk about project management basics – because controlling your
vendor also means keeping the project on track. In the project world,
we say *quality*, *price*, and *time* are the three variables, and you can't
maximize all three at once. If you want it fast and cheap, quality will
suffer. If you want top quality and low cost, it will take longer. If you
need high quality in a short time, it's going to cost more in resources.
No agency can break this physics of project management, no matter
what they claim. So be realistic and prioritize what matters most to
you. I always ask clients to rank what's more important: hitting a date,
staying under $X budget, or achieving a certain scope/quality level –
because inevitably we'll need to make trade-offs. If an agency promises
"we'll do everything you want, in half the time, and it'll cost less than
the other guys," be extremely skeptical. They're most likely over-
promising.

There's a fun analogy in the software planning world: *nine women
can't make deliver a baby in one month*. Some things just take the time
they take and adding more people doesn't help. Adding more people
to a project doesn't always speed it up; in fact it can slow things due to
coordination overhead (Brooks's Law in software development). I say
this to set your expectations: when planning with your agency, push
for a realistic timeline, not an overly optimistic one that will almost
surely slip. If you absolutely have a hard deadline (say, a Black Friday
launch or a CEO-mandated date), make sure the scope is adjusted to fit
that timeline – again, trade-offs. Good agencies will be honest about
how long things will likely take once they understand requirements. If
everyone is honest up front, you can avoid the common scenario of "I
want it yesterday and I want it perfect" leading to disappointment.

A tactic I highly recommend for larger projects is a **phased or
sprint-based approach**. Instead of one giant "launch" six or nine
months out, break the work into phases that deliver some value earlier.
This could be Phase 1: launch core eCommerce features, Phase 2: add

advanced personalization, etc. Phases give you proof points – you can start generating ROI earlier – and they also let you **test the partnership**. If Phase 1 goes well, great, continue. If it was rocky, you have an opportunity to course-correct or even reconsider the relationship before committing to further phases. I often treat the first 90 days of an engagement as a trial period (even if not formally stated). You as a client should assess: did the agency deliver on time? Was the quality good? How was communication? The agency is evaluating you too, by the way – are you responsive, decisive, reasonable? If both sides are clicking, awesome, keep going (maybe even expand the partnership). If not, it's a manageable point to part ways or adjust terms, rather than suffering through a long painful project that fizzles out.

Maintain a Sense of Urgency & Ownership of the Timeline. Earlier I mentioned not letting weeks slip by with no progress. Let me hammer that home: always have **crystal-clear next steps and owners after every meeting**. In my consulting, I sometimes join calls between a client and their incumbent agency just to observe. I'm amazed how often meetings end with "Okay, we'll regroup soon" and no one saying who will do what by when. Then a month passes and nothing's changed – except now the deadline is a month closer, so everyone panics. Don't fall into this trap. If you have an internal project manager or lead, task them with driving this rigor. If not, hold the agency's project manager accountable to do it. Every status call should end with a rundown like, "Alright, Action Items: John will provide the updated pricing spreadsheet by Tuesday, Agency will deliver the revised homepage design by Friday, and we'll meet again next Monday to review. Agreed?" Document it in an email if needed. It sounds basic, but trust me, it's these simple habits that keep projects on schedule.

Also, instill a bit of urgency in the culture of the project. I'm not talking about stressful crunch vibes, but a shared understanding that *time is money* in eCommerce. **Every day that slips is a day your new solution isn't live generating revenue**. A six-week delay might mean tens or hundreds of thousands in lost sales opportunity or increased project cost. When everyone realizes that, people tend to respond faster and push a little harder to hit dates. As the client, set the tone by promptly doing your part – whether it's approvals, content delivery, or testing. That motivates the agency to do the same. If you drag your feet on feedback for two weeks, you can't expect the agency to magically still hit the original timeline. So model the urgency you want to see.

Finally, don't be shy about revisiting priorities if something changes. It's better to adjust scope or push out a secondary feature than to let the whole project slip because you're trying to do too much. This is again where transparency helps: if you're falling behind, talk about it. Re-plan together, rather than suffering in silence until a deadline is missed. A consistent theme here is **active management** – a great vendor relationship is like a dance, both sides need to stay in step and communicate.

WHEN IT'S TIME TO WALK AWAY

Despite your best efforts in selecting a strong partner and managing the relationship well, not every vendor is a forever partner. Situations change. Maybe the agency grew and their service level declined, or your favorite account manager left. Maybe your platform can't support your growth anymore. How do you know when it's time to part ways or "re-platform" to a new solution? Let's go over some **red flags and exit strategies**. As someone who's been on **both** sides of these breakups (I've been the incoming partner replacing an incumbent, and I've unfortunately been the incumbent being replaced), I have some candid perspective on this.

Red Flags – Signs Your Provider Is Holding You Back: In an earlier chapter we covered monitoring KPIs and QBR metrics; those should give you objective signals. Here are some big ones that often trigger the "should we switch?" question:

Frequent Downtime or Instability: If your site crashes or goes down regularly, something is wrong – either with your platform or your agency's ability to maintain it. This is eCommerce, not a hobby site; uptime is money. Don't accept excuses like "well, eCommerce sites have issues sometimes." Yes, things happen, but if it's a pattern, your partner might be out of their depth. I often say in these cases: *either the technology isn't right for your scale, or the team supporting it isn't doing a good job.* Both are solvable by changing one or the other. Also, instability erodes customer trust – shoppers won't return if your site is unreliable, and you can even get dinged in SEO rankings if Google finds your site unresponsive. So this is an urgent red flag.

Can't Handle Growth or Peak Load: Similar to downtime but slightly different angle – maybe the site stays up but performs terribly under high load, or you constantly worry it *might* go down if traffic spikes. If your current platform or agency panics every holiday season, pulling all-nighters just to keep servers running, that's not sustainable. As we discussed, top-tier platform providers plan ahead for peak events. If yours isn't, or your agency doesn't even know how to scale the site for growth, you will be handcuffed. The whole point of investing in eCommerce is to grow revenue without hitting a ceiling; an inability to scale is a loud signal that you may need a more robust platform or a more experienced partner.

Slow to Innovate: ECommerce evolves fast – new tech, new consumer expectations. If you find your platform lacks capabilities for modern trends (like a headless front-end, AI integration, personalization, etc.), or your agency is dismissive of new approaches, you risk falling behind competitors. A recent example: we were early in exploring how AI (like ChatGPT) could streamline operations. One prospective client's agency had never even considered using AI for, say, automating customer service FAQs or generating basic product descriptions. We prototyped an integration that saved this client thousands per month in support costs. The old agency's lack of curiosity would have cost the client real money in the long run. So, ask yourself: has your partner brought you any fresh ideas lately? Are they proactively suggesting improvements, or do they seem to be coasting? If the latter, maybe they've grown complacent. An innovative, growth-focused business eventually outgrows a stagnant partner.

Roadmap Stagnation: If two quarters in a row go by and none of the new features or improvements you planned are getting delivered, it's time to take a hard look. It could be internal issues, but if your team is functioning and the hold-up is on the vendor side (missed estimates, resource churn, etc.), you might need a change. Look at trend lines: one off-quarter is understandable, but if you consistently can't hit roadmap goals due to your agency's execution, the writing's on the wall.

Chronic Red Metrics: Earlier we set up a scorecard. If key metrics (like those velocity or quality indicators) are red for *three consecutive quarters* despite conversations and attempts to address them, that's a pretty clear sign things aren't working. For instance, if your bug count is high release after release, or speed of delivery remains way below target quarter over quarter, the team might just not be capable of meeting your needs. It's tough to admit, but better than limping along.

Surprise Bills and Scope Creep: If you keep getting unexpected invoices for things you didn't anticipate, or the project cost keeps blowing past estimates with little warning, trust is eroding. One surprise invoice might be a misunderstanding; repeat surprises indicate either poor planning or an exploitative practice. Either way, not good. Similarly, if the scope is endlessly creeping (and not because *you* keep requesting new features) but because the agency can't deliver within original scope, it may be time to cut losses. Remember the earlier advice: transparency on finances. If despite that you still feel nickel-and-dimed, start evaluating alternatives.

Security Neglect: Here's a non-negotiable: Are critical security patches and updates applied to your platform promptly? I generally say within a week of release for high-severity patches is a reasonable expectation. If your agency regularly lets weeks or months pass on important security updates, they're not prioritizing your business's safety. Security should be table stakes. A lax approach here is not only a reason to switch – it's a liability for your whole company.

Dread and Poor Communication: Lastly, trust your gut and emotions. Do you *dread* meetings with your agency? Does the thought of deploying a new change fill you with anxiety because you fear it will go wrong? If every interaction is stressful or combative, it might simply be a broken relationship. Sometimes, through no specific "fault" of either side, the chemistry just isn't there or has deteriorated. Business partnerships are like marriages in some ways – if you've tried counseling (i.e. open conversations, maybe escalation to higher management) and it's still miserable, a separation might be healthiest. Life's too short and eCommerce too challenging to be stuck in a bad partnership.

So, say you've spotted one or more of those red flags and decided, *"Yeah…we need to make a change."* How do you exit gracefully and set yourself up for success with the next partner? Here's the playbook I recommend, much of which we include in our E-Commerce Growth downloadable resources (like the re-platforming guide mentioned later):

Plan the Transition – Don't "Rip and Replace" Overnight. The ideal scenario is a **managed, overlap transition** over perhaps 90 days. Check your contract for any notice period or termination clause – you may need to give 30 days notice or pay for a final month, etc.. Honor those terms to avoid legal spats. I actually advocate being generous here: if you can afford it, tell the outgoing agency, *"We'll keep you on retainer for 2–3 months during the transition."* This keeps them cooperative and engaged (since they're still getting paid) while the new agency ramps up. It's an investment in a smooth handover. Outline a knowledge transfer plan with them. You want their team to document or explain the key parts of the system to your incoming team. One common pitfall: brands wait until they're leaving to suddenly demand tons of documentation from the old agency. If documentation wasn't in the contract, they might rightfully say, "sure, at our hourly rate… that'll be 100 hours of work" – not what you want to hear at the end. Ideally, you would have been investing in documentation all along (many don't, I get it). If you haven't, then at least have the outgoing team walk the new team through the code and architecture. Have them provide whatever existing docs they do have (deployment procedures, architectural diagrams, etc.).

Next, **audit all access and assets**. This is where your foresight in "owning the keys" pays off. Make a checklist: Source code repositories, hosting accounts, DNS records, SSL certificates, third-party integrations, admin passwords, analytics accounts, email sending domains – all of it. Ensure the outgoing agency has handed over or you have changed credentials to cut their access at the appropriate time. You don't want any loose ends where an old developer still has write access to your code or an old PM can log in to your Google Analytics. It's not about distrust (hopefully you parted on okay terms); it's just good security hygiene. Many breaches and mishaps happen during transitions, so be thorough here.

Then, **set a firm cut-over date** for the new agency or platform to take full control. During the transition period, it's fine to have both old and new agencies involved – perhaps the old team is still handling urgent bug fixes while the new team learns the ropes. But *do not* run two different agencies doing changes on the same codebase at the same time in parallel without coordination. That's a recipe for confusion ("crossing the streams," as I call it, like in Ghostbusters – just don't do it). Pick a date where after that, only the new agency will deploy code. Leading up to it, maybe freeze any non-critical changes so the new team can start with a stable baseline. Communicate this date to all stakeholders. It helps avoid the scenario of "I told the old agency about that request" – "oh, I thought the new agency was doing it" – and things fall through the cracks.

A note on re-platforming (switching eCommerce platforms) if that's part of your change: that's a bigger topic, but in brief, you'll want to run a thorough requirements and vendor selection process (similar to how we picked an agency) for the new platform. Analyze what your business will need for the next 3-5 years, not just today. The download-able **Replatforming Guide** (see the end of this book) can assist in eval-uating platforms and planning the migration step by step. Often a platform switch and agency switch go hand-in-hand, but not always – sometimes you stick with the software but change the service partner, or vice versa. Either way, approach it systematically: do discovery, migrate data carefully, and time it in a slow season if possible. The worst is rushing a migration due to a vendor breakup – try to avoid urgency by planning ahead when you first sense things are going south.

One more thing: **keep it professional and courteous** with the outgoing partner. The eCommerce world is surprisingly small. You never know when paths will cross again. Plus, you want their full cooperation during transition. So, as frustrating as things may have been, leave emotions out of it. State the reasons for moving on if you feel it could help them improve (who knows, maybe they'll take the feedback to heart for their next client), but don't burn bridges unneces-sarily. At the end of the day, this is your business decision to find the best fit, and any reasonable agency will accept that.

Lastly, consider doing a **post-mortem** after the transition: What did we learn from this failed partnership? Were there warning signs we missed early on? How will we ensure the next relationship starts differently? This book isn't about dwelling on the negative, but continuous improvement means learning from mistakes. Perhaps you realize you should have instituted QBRs earlier, or maybe the vendor's size wasn't right for you (too big or too small). Take those lessons into your next 90-day execution sprint (coming up next!) as you implement changes.

Staying too long with a bad vendor is costly. I've had merchants come to me after enduring an underperforming agency for *years*, and the regret is palpable. They lost market share, their team morale suffered working with poor systems, and they essentially paid high fees for subpar results. Don't let inertia or fear of change keep you stuck. Yes, switching has a short-term cost and risk, but the upside of a better partnership – faster growth, happier customers, saner work days – is well worth it. Remember, your competition is always one click away for your customers. If you're not continually improving your eCommerce experience, somebody else is, and they'll happily take those clicks (and dollars) off your plate. Sometimes the bravest (and smartest) thing a leader can do is say "enough is enough" and make a change.

We've covered a lot in this chapter: how to choose the right eCommerce partner and how to actively manage that relationship for maximum success. From vetting their expertise and culture fit, to setting up governance processes like QBRs and scorecards, to keeping control of your technology and knowing when to cut ties – these are the ingredients of *vendor control* that keep your growth engine humming. It's not easy, I know. It requires time, frank conversations, and the courage to stick to high standards. But the reward is huge: an eCommerce operation that runs smoothly, adapts quickly, and delivers results year after year.

As we wrap up this section of the Ecommerce Growth Playbook, take a moment to reflect on your current agency and vendor relationships. Are you applying these principles? Are you truly treating them as partners – and are they acting like one? What immediate adjustments can you make (maybe scheduling that first QBR, or doing an access audit, or having that heart-to-heart with your account manager)? Jot down a few actions.

Now, here's where the *pragmatic* part comes in. In the next part of the book, we're not just going to talk about it – we're going to **do** it. Up next is a **90-Day Execution Sprint** plan. Consider it your roadmap for the next quarter, translating all these strategies into concrete action steps. Whether it's implementing a new KPI scorecard, initiating a re-platforming project, or conducting a vendor review, we'll outline how to tackle it in 90 days to build serious momentum. Think of it as your coaching program for the next three months – I'll be there in spirit, pushing you to follow through. So, get ready to roll up your sleeves and dive into execution mode. By the end of that sprint, you'll be amazed at how much progress you can make by focusing on the right things with the right partners.

5.1: YOUR 90-DAY ACTION PLAYBOOK

T alk is cheap; shipping wins revenue. But big plans stall when real life walks in. That's why this book closes with a tactical sprint: three 30-day blocks that stack fast wins into long-term momentum. Work the plan in order—or, if a fire's burning, skip ahead to the block that solves it first. Finish the 90 days and you'll have removed friction, tuned your tech, hard-wired accountability, and primed your team and your ecommerce business for continuous growth.

BLOCK 1 — DAYS 1-30: PLUG THE BLEEDS & BUILD THE ENGINE
Goal

Stop silent revenue leaks and lay a foundation that the next 60 days can scale. Here's the specific parts of specific chapters to focus on.

Chapter 4.3 - Baseline metrics & scorecard. You can't improve what you can't see.

Chapter 2.1 - Friction audit & quick fixes. Fastest path to instant conversion lift.

Chapter 2.1 - UX, search, checkout wins. Compounds the friction fixes.

Chapter 3.1 - Tech-debt triage & first patch. Stability before scale.

Chapter 4.1 - Pilot Growth Pod + Scorecard huddle. Gives the sprint an owner and a heartbeat.

Week-by-Week Play

Week 1 — Baseline & Ownership

Record current CR, AOV, 60-day repeat rate, deployment frequency (Ch 1.1, 4.3).

Name one KPI owner for each.

Drop the five numbers into a Google Doc Scorecard and schedule a 15-min weekly huddle.

Week 2 — Friction Squad & Mobile First

Run the full mobile journey yourself (Ch 2.1).

List every speed bump: load lag, pop-ups, broken autofill.

Remove or defer one third-party script; compress hero images; test again.

Week 3 — Search & Checkout Boost

Pull search logs; surface the top ten "no results" queries (Ch 2.2).

Add search synonyms or fix typos; rerun test.

Slash checkout fields to ≤ 6 and turn on guest checkout + Apple/Google Pay.

Week 4 — Tech-Debt Blitz

Static-scan the codebase; find one high-impact, low-complexity fix (Ch 3.1).

Deploy it behind a feature flag.

Track page-speed or bug rate deltas.

Milestones to Hit by Day 30

Mobile LCP <2.5 s

Qualified Sessions +5 %

Checkout Fields reduced to ≤ 6

Deployment Frequency - once every 2 weeks

End-of-Block Checklist *(paste to your wall)*
- Scorecard live with 5 metrics & owners
- Weekly 15-min KPI huddle scheduled
- Friction Squad chartered & backlog created
- Mobile LCP < 2.5 s on home + PDP
- Guest checkout & wallets enabled
- One tech-debt item fixed behind a flag
- Celebrate first win publicly

Block 2 — Days 31-60: Accelerate & Align

Goal

Scale gains across channels, personalize without creepiness, and lock change rhythms before fatigue hits. Here's the specific parts of specific chapters to focus on.

Chapter 2.3 - Personalization Ladder Levels 1–3. Increases AOV & repeat without major build.

Chapter 2.4 - Omnichannel Pilot (BOPIS or Ship-from-Store). Captures margin once speed is fixed.

Chapter 3.3 - Data Plumbing Walk Stage. Prevents oversells as channels merge.

Chapter 3.4 - Core Web Vitals culture. Protects gains in search and UX.

Chapter 4.2 - Change Flywheel + Energy pulse. Keeps team batteries charged.

Week-by-Week Play

Week 5 — Personalization Audit

Run the five-level ladder self-test (Ch 2.3).

Choose the weakest rung—likely Level 1 Recognition.

Turn on merge-tag greetings across email & account pages.

Week 6 — Search & Merch Contextual Wins

Enable "Recommended for you" blocks on top-10 SKUs.

Track upsell conversion rate; target 5 %.

Add "People also bought" to cart drawer.

Week 7 — Omnichannel MVP Kickoff

(For brands that don't have store locations, use this week for enhancing a different channel, such as marketplaces.)

Pick one pilot store for BOPIS (Ch 2.4).

Audit top-100 SKUs; hit 95 % inventory accuracy.

Brief store manager; set a pickup-ready KPI of < 2 hrs.

Week 8 — Plumbing Upgrade

Move ERP → site inventory sync from nightly to hourly (Ch 3.3 'Walk').

Set failure alerts in Slack/Teams.

Create anomaly guardrail: alert if a product's price drops 50 %+ suddenly.

Week 9 — Speed Culture Embed

Add CWV pass/fail to CI; block deploys exceeding performance budgets (Ch 3.4).

Set real-user monitoring alerts at p75 LCP > 2.5 s.

Week 10 — Energy Pulse & Showcase

Run a 3-question eNPS pulse (Ch 4.2).

If score < 40, cut low-value meetings for a week.

Host Friday demo: show BOPIS pilot orders, upsell data, faster sync.

. . .

Milestones to Hit by Day 60

    60-Day Repeat Rate - baseline +2 pts

    Upsell CVR - baseline +5 %

    Pickup Ready Time < 2 hrs

    Inventory Sync Lag <= 15 min

End-of-Block Checklist

- Personalized greetings live; open + click up
- Contextual upsells on 10 SKUs; >5 % CVR
- BOPIS pilot live in one store
- Inventory sync moved to ≤15 min
- CWV budgets enforced in CI
- eNPS pulse ≥40 or actions scheduled
- Demo recorded & shared company-wide

Block 3 — Days 61-90: Compound & Future-Proof
Goal

Lock in flywheel momentum: decouple your tech, tighten vendor control, and set a quarterly growth cadence. Here's the specific parts of specific chapters to focus on.

Chapter 3.2 - Composable Pilot or Integration Hub. Future-proofs scale after initial fixes.

Chapter 3.1 - Tech-Debt Blitz round 2. Keeps engine lean.

Chapter 2.3 - Higher-Level Personalization + AI (Level 4–5). Drives loyalty.

Chapter 4.4 - Vendor Scorecard & QBR. Ensures partners don't slow the flywheel.

Epilogue (6.1) - 90-Day Continuous-Improvement Loop. Sets permanent rhythm.

Week-by-Week Play

Week 11 — Composable Litmus & Path
Score the five-question test (Ch 3.2).
Choose either Front-End-First or Integration-Hub path.
Spin up sandbox for microservice (search, CMS, or checkout) next door to the monolith.

Week 12 — Vendor Audit & Ownership Fix
Inventory all licenses, hosting, repos (Ch 4.4).
Transfer any still in vendor names.
Create a five-KPI vendor scorecard: velocity, bugs, uptime, budget variance, NPS.

Week 13 — Predictive & Proactive Personalization
Launch a refill reminder or predictive bundle email (Ch 2.3 Level 4).
Auto-alert for shipping delays with perk credit (Level 5).
Measure opt-in %. Goal 8–12 %.

Week 14 — Tech-Debt Blitz # 2 & Speed Retest
Fix next high-impact debt item.
Re-run full Lighthouse + real-user CWV tests; confirm still green.

Week 15 — Integration Go-Live
Roll microservice to 10 % traffic behind feature flag.
Monitor API error < 0.2 %.
If green for 48 hrs, roll to 100 %.

Week 16 — QBR & 100-Day Plan
Hold first Quarterly Business Review with agency & platform vendor (Ch 4.4).
Review Scorecard trends from Blocks 1-3.

Set next quarter's Rocks: two Hero KPIs to move, one tech-debt cluster to kill, one growth experiment.

Milestones to Hit by Day 90
Deployment Frequency to 1 per week
60-Day Repeat Rate +4 pts
Vendor Scorecard Reds 0
API Error Rate < 0.2 %
Tech-Debt Backlog: baseline –20 %

End-of-Block Checklist
- Microservice live & stable
- All licenses / hosting in your name
- Vendor scorecard in QBR deck
- Predictive personalization campaign
- CWV still green after new code
- Next-quarter Rocks & owners chosen

6.1: MOMENTUM - KEEPING THE FLYWHEEL SPINNING

You've finished the book, and the 90 day growth plan. You've slashed friction, sped up pages, fixed data leaks, empowered a Growth Pod, personalized without creepiness, tightened vendor control, and shipped weekly. That's not a project; it's a **flywheel**. What now? Keep it spinning by renewing the Scorecard every quarter, rolling tech-debt blitzes into every sprint, and graduating KPIs once they stay green.

This playbook isn't meant to be a checklist to tick once. It's a loop to run forever—because the only competitive moat left in ecommerce is how fast you learn, ship, and improve.

You've just hiked through 100,000 words of frameworks, checklists, and stories—more terrain than most ecommerce executives cover in an entire career. If you stuck with me, you already know the biggest secret in mid-market e-commerce: growth isn't magic. It's continually working to improve on three levers you control every single week.

LEVER 1 — CUSTOMER OBSESSION

Every dollar starts or dies with each individual shopper's micro-experience. Remove friction, personalize without creepiness, deliver two-click refunds, keep LCP under 2.5 seconds, and your flywheel accelerates on its own. Amazon doesn't win because it's big; it wins because every feature proposal begins with the question, "Does this make life easier for the buyer?" Adopt that approach and you'll squeeze pure margin out of the same ad spend and traffic you already have.

LEVER 2 — SCALABLE ENGINE

Faster pages, real-time inventory, composable services, automated tests, weekly deploys—this is how IT drives ecommerce revenue. When a TikTok mention slams a million page views at your site, marketing celebrates, Ops ships, and finance smiles **only** if your ecommerce tech stack thrives under load. Otherwise everyone spends Sunday firefighting while shoppers slide to your competitor. Treat tech debt like interest on a bad credit card and pay it down before it swallows your budget!

LEVER 3 — LEADERSHIP & ORG DESIGN

Tools don't execute; people do. Cross-functional pods with a single Product Owner out-run silos every time. Weekly Scorecards keep the truth visible, and EOS Level-10 meetings cut drama. Culture isn't kombucha and beanbags; it's a team that knows exactly who owns a KPI and feels safe admitting a red number before it becomes a headline.

A STORY ABOUT MOMENTUM

Six months ago a $30 million B2B distributor called us in a panic. Their site crashed every Cyber Monday and they'd just lost their lone e-commerce manager to burnout. We didn't sell them a replatform; we sold them momentum.

Week 1: We ran the five-number Scorecard workshop. CR was 1.8 %, AOV $112, repeat rate 14 %, deploy frequency once a month, and mobile LCP an embarrassing 4.6 seconds.

Week 2: One developer removed two abandoned scripts and flipped Cloudflare on. LCP dropped to 2.8 s overnight. The team saw green for the first time in a year.

Month 2: A freshly minted Growth Pod shipped guest checkout and Apple Pay. Conversion popped to 2.2 %. Engineering finally believed weekly deploys wouldn't explode the server.

Month 3: We integrated their creaky ERP to send stock counts hourly instead of nightly. Oversells fell 88 %. Support tickets shrank; morale inflated.

By Day 90 their Scorecard lit up like a Christmas tree: CR 2.6 %, AOV $225, repeat 17 %, deploys weekly, mobile LCP 2.1 s. Same traffic, same ad spend—just fewer leaks and more speed. The CFO didn't need another pitch deck; he needed proof. Momentum sold itself.

I'm not sharing that to brag. I'm sharing it because there was nothing exotic in that engagement. The team used the exact checklists you now hold. You just have to do the work.

Your Next Step—Now, Not Tomorrow

Pick One Lever.

Which statistic above made you wince? Start there. If bounce rate screams, friction is the lever. If deploys scare your engineers, fix the engine. If everyone dodges ownership, reshuffle the org. Circle it in red.

Set a 15-Minute Timer.

Open your analytics, your Jira board, or your org chart—whatever relates to that lever. Identify one number that proves the pain. Write it on a sticky note; slap it on your monitor.

Book the First Huddle.

Invite every stakeholder to a 15-minute call. Agenda: show the number, ask "What's the smallest experiment we can ship next week to move it?" Nothing more. End the meeting.

Ship the Experiment.

Kill a script, add Apple Pay, compress an image, rewrite a modal—anything you can launch inside five workdays. Perfection is paralysis.

. . .

Measure and Broadcast.

Did the number move? Screenshot the metric, drop it in Slack, tag the team, throw in a meme. Public wins fuel private grit.

Do that once and you'll want to do it again. Do it four times and you've built a habit. Do it twelve weeks in a row and you won't recognize your dashboard—or your culture.

A Quick Word on Resistance

You'll hear three objections:

"We don't have bandwidth." Translation: priorities are fuzzy— Scorecard them.

"We need a bigger budget first." False. Most of the wins in this book cost much less than you'd expect. Reach out to me if someone's telling me you need a six figure budget to fix these items, because you don't.

"Let's wait until after peak season." Growth won't wait; your competition won't either. Small changes today beat heroic launches next year.

I have ADHD (and I feel like most ecommerce teams somehow do too); shiny ideas outnumber Tuesday afternoons. The only hack that works is motion. Micro-actions destroy overthinking. The playbook is purposely built from bite-size tasks because that's the only way **I** stay on track. Trust the system.

An Invitation

I want your war stories. Email me the moment you ship your first micro-win. Tell me the metric, the tweak, and the impact. I answer every note, even if it takes a midnight coffee. And I'd love to have you come on the Commerce Today podcast and share your story with the broader ecommerce community.

If the entire book felt like gulping from a fire hose, start with the **90-Day Action Playbook** in the appendix. Follow it line by line. At

Day 91 ping me with results—good, bad, or ugly. I'll trade you a 30-minute coaching call for the debrief. Real feedback sharpens the next edition and sharpens my agency's advice.

Final Charge

Growth isn't a mystery; it's a choice repeated daily. The levers are right there in front of you:

Obsess over the customer until friction is eliminated.

Build a scalable engine so speed and stability stay green when marketing finds its next traffic geyser.

Design a culture that ships, learns, and ships again—because yesterday's wins spoil fast.

Close this book, fire up your timer, and start on one metric today. Momentum is the only moat you can actually control. Go build it.

I NEED YOUR HELP!

If you found the information in this book helpful, please take a moment to leave a review on Amazon.

Like any other product on Amazon, reviews determine if the product page highlighting this book will be displayed when people are searching for related topics. So, every review really does help.

Please also spread the word to other ecommerce professionals that would find this book useful.

RESOURCES

For the Creatuity resources mentioned in each chapter, visit https://creatuity.com/playbook and for third-party resources and sources, see the per-chapter list that follows.

Chapter 1.1 — Why "Good Enough" Is a Silent Revenue Tax

Site & Analytics Basics

PageSpeed Insights (https://developers.google.com/web/tools)
WebPageTest (https://webpagetest.org)
Google Analytics 4 (https://analytics.google.com)

Books & Frameworks

Good to Great – Jim Collins (https://jimcollins.com)
Traction: Get a Grip on Your Business – Gino Wickman → core EOS text (https://www.eosworldwide.com)

Podcasts

Commerce Today (https://commercetoday.fm)
eCommerce Fuel (https://www.ecommercefuel.com/podcast)

Chapter 2.1 — The Cost of Friction

Speed & UX Tools
Core Web Vitals (https://pagespeed.web.dev/)
Hotjar Session Replay (https://hotjar.com)
Lucky Orange Funnels (https://luckyorange.com)

Returns & Trust
Loop Returns (https://loopreturns.com)
Narvar Track & Return (https://narvar.com)

Reading
The Best Service Is No Service – Frei & Morriss

Chapter 2.2 — Experience Engineering: UX, Search, Checkout

UX & Accessibility
Axe DevTools (https://deque.com/axe)
Stark color-contrast plugin (https://www.getstark.co)

Search & Merch
Algolia (https://algolia.com)
Typesense (https://typesense.org) – open-source option

Checkout & Payments
Stripe Checkout (https://stripe.com/payments/checkout)
Shop Pay (https://shop.app/pay)
Klarna BNPL (https://klarna.com)

Books
Don't Make Me Think – Steve Krug

Chapter 2.3 — Personalization Without Creepiness
Platforms
Klaviyo Flows (https://klaviyo.com)
dotdigital Customer Data (https://dotdigital.com)

Segment CDP (https://segment.com)

AI & Recommendation Engines

Google Retail Recommendations AI (https://cloud.google.com/recommendations)

Recombee (https://www.recombee.com)

Inspirational Reads

Be Our Guest – Disney Institute (https://books.disney.com)

Chapter 2.4 — Omnichannel That Actually Works

Inventory & OMS

Airtable Stock Tracker MVP (https://airtable.com)

ShipBob WMS/API (https://shipbob.com)

Curbside/BOPIS Tech

Alert Innovation Light Speed Pick-Up (https://alertinnovation.com)

Scandit Smart Data Capture (https://scandit.com)

Podcasts

Retail Gets Real (https://nrf.com/retailgetsreal)

Chapter 3.1 — Tech Debt: The Silent Killer of Margin
Scanners & CI

SonarQube (https://sonarsource.com)

GitHub Actions CI (https://github.com/features/actions)

LaunchDarkly Feature Flags (https://launchdarkly.com)

Reading

Accelerate – Forsgren, Humble & Kim (https://itrevolution.com/accelerate)

Chapter 3.2 — Composable vs Monolith
Composable Stack Starters

Contentful CMS (https://contentful.com)
Netlify Edge Functions (https://netlify.com)

Integration / iPaaS
iPaaS (https://ipaas.com)
Celigo (https://celigo.com)
MuleSoft (https://mulesoft.com)

Industry Frameworks
MACH Alliance Guides (https://machalliance.org)

Chapter 3.3 — Data Plumbing: ERP, PIM, Real-Time Inventory

PIM Options
Akeneo PIM (https://akeneo.com)
Pimcore (https://pimcore.com/en/)
Plytix PIM (https://plytix.com)

ERP & Connectors
NetSuite (https://netsuite.com)
Acumatica (https://acumatica.com)

Monitoring
Datadog Integration Errors (https://datadoghq.com)

Chapter 3.4 — Site Speed & Core Web Vitals

Perf-Ops
New Relic Browser (https://newrelic.com)
Cloudflare CDN & RUM (https://cloudflare.com)
Hyvä Theme for Magento (https://hyva.io)

Load Testing
k6 Cloud (https://k6.io)
SpeedCurve (https://speedcurve.com)

Chapter 4.1 — Org Design for E-commerce 3.0
People & Process
EOS Level-10 Meeting™ Agenda (https://www.eosworld-wide.com)

Atlassian Jira Software (https://www.atlassian.com/software/jira)

Loom Async Demos (https://loom.com)

Hiring Culture Reads
Who – Smart & Street (https://whothebook.com)

Chapter 4.2 — Change Management When Everyone's Tired

Change Flywheel Helpers
Mural Sprint Boards (https://mural.co)

Officevibe eNPS Pulse (https://officevibe.com)

Pomodoro Focus App – Focus To-Do (https://focustodo.cn)

Books
Switch – Chip & Dan Heath (https://heathbrothers.com/books/switch)

Atomic Habits – James Clear (https://jamesclear.com/atomic-habits)

Chapter 4.3 — Metrics That Matter

Dashboards
Looker Studio (https://lookerstudio.google.com)

ShopifyQL Notebooks (https://shopify.dev/docs/custom-store fronts/shopifyql-notebooks)

Adobe Commerce Business Intelligence (https://experienceleague.adobe.com/en/docs/commerce-business-intelligence/mbi/getting-started)

Scorecard Inspiration
Tableau KPI Playbook (https://tableau.com/learn)

Chapter 4.4 — Partner Selection & Vendor Control

Contract & Security Must-Haves
1Password Shared Vaults (https://1password.com)
OSCR Project Risk Register (https://oscr-template.io)

Always-Useful Cross-Chapter Extras

Creatuity Tools & Services
Commerce Today Podcast Archive (https://commercetoday.fm)

Widely Respected Reads
Measure What Matters – John Doerr (OKRs) (https://measurewhat-matters.com)

The Phoenix Project – Kim, Behr & Spafford (DevOps Fiction) (https://itrevolution.com/phoenixproject)

INDEX

ABOUT THE AUTHOR

I've spent over two decades turning creaky online stores into revenue engines. As founder and CEO of **Creatuity (creatuity.com)**—the Texas-born, global-powered ecommerce agency—I've led replatforms, speed rescues, and nine-figure growth programs for B2B suppliers, farm-and-home chains, and pure-play DTC disruptors. My team's work has cut load times by 70 %, doubled conversion rates, and helped mid-market brands punch well above their weight.

Every week on the **Commerce Today** podcast I unpack those wins, the stumbles that taught us more, and the frameworks that made both repeatable. If you like the book's no-fluff style, you'll love the episodes —they're the live-fire lab where these playbooks were forged.

Away from the mic I coach ecommerce leaders on scorecards, tech-debt blitzes, and vendor governance that actually protects profit. I nerd out on Core Web Vitals, but my real obsession is helping teams ship faster and celebrate bigger wins.

Let's compare notes. Got a metric that moved because of this playbook? Or a mess you can't untangle? Reach out—I answer every thoughtful message. First coffee's on me next time I'm in your city.
- **Email:** josh@creatuity.com
- **LinkedIn:** https://linkedin.com/in/joshuawarren
- **Podcast:** https://commercetoday.fm
- **Agency:** https://creatuity.com

If you're looking for the Creatuity resources mentioned throughout the book, you can find them at https://creatuity.com/playbook